普通高等教育"十一五"国家级规划教材
全国高等医药院校药学类实验双语教材

电工电子学实验与指导

Experiment and Guide for Electrical Engineering and Electronics

主　编　张　琪
编　者　(以姓氏笔画为序)
　　　　张　利　张爱平
　　　　张　琪　钟　宁

中国医药科技出版社

内 容 提 要

　　本教材是全国高等医药院校药学类实验双语教材之一。全书采用中英文对照形式，根据电工技术课程及电子技术课程教学基本要求并结合现有的实验设备条件编写。本书包括 15 个实验，涉及电路、电机及控制、模拟电子技术等部分。除了基本实验外，还增加了设计性实验的比重。书中通过明确预习要求、实验报告要求等手段注重了能力的培养，适应现代化的教学要求。

　　本书可供高等院校药学、化学类专业工科学生使用，也可以作为电类专业教学及电子工程技术人员的参考书。

图书在版编目（CIP）数据

电工电子学实验与指导：汉英对照/张琪主编 . —北京：中国医药科技出版社，2007.5

普通高等教育"十一五"国家级规划教材 . 全国高等医药院校药学类实验双语教材

ISBN 978 - 7 - 5067 - 3642 - 8

Ⅰ . 电⋯　Ⅱ . 张⋯　Ⅲ .①电工技术—实验—双语教学—高等学校—教材—汉、英②电子技术—实验—双语教学—高等学校—教材—汉、英　Ⅳ . TM - 33　TN - 33

中国版本图书馆 CIP 数据核字（2007）第 075928 号

美术编辑　陈君杞
责任校对　张学军
版式设计　郭小平

出版　中国医药科技出版社
地址　北京市海淀区文慧园北路甲 22 号
邮编　100082
电话　发行：010- 62227427　邮购：010- 62236938
网址　www.cmstp.com
规格　787×1092mm $\frac{1}{16}$
印张　11
字数　254 千字
版次　2007 年 5 月第 1 版
印次　2019 年 6 月第 3 次印刷
印刷　北京市密东印刷有限公司
经销　全国各地新华书店
书号　ISBN 978-7-5067-3642-8
定价　20. 00 元
本社图书如存在印装质量问题请与本社联系调换

出 版 说 明

　　全国高等医药院校药学类专业规划教材是目前国内体系最完整、专业覆盖最[全]面、作者队伍最权威的药学类教材。随着我国药学教育事业的快速发展，药学及[相]关专业办学规模和水平的不断扩大和提高，课程设置的不断更新，对药学类教材[的]质量提出了更高的要求。

　　全国高等医药院校药学类规划教材编写委员会在调查和总结上轮药学类规划[教]材质量和使用情况的基础上，经过审议和规划，组织中国药科大学、沈阳药科大学、广东药学院、北京大学药学院、复旦大学药学院、四川大学华西药学院、北京中[医]药大学、西安交通大学医学院、华中科技大学同济药学院、山东大学药学院、山[东]医科大学药学院、第二军医大学药学院、山东中医药大学、上海中医药大学和江[西]中医学院等数十所院校的教师共同进行药学类第三轮规划教材的编写修订工作。

　　药学类第三轮规划教材的编写修订，坚持紧扣药学类专业本科教育培养目标，参考执业药师资格准入标准，强调药学特色鲜明，体现现代医药科技水平，进一[步]提高教材水平和质量。同时，针对学生自学、复习、考试等需要，紧扣主干教材[内]容，新编了相应的学习指导与习题集等配套教材。

　　本套教材由中国医药科技出版社出版，供全国高等医药院校药学类及相关专[业]使用。其中包括理论课教材 82 种，实验课教材 38 种，配套教材 10 种，其中有[几]种入选普通高等教育"十一五"国家级规划教材。

全国高等医药院校药学类规划教材

编写委员会

2009 年 8 月 1 日

序

实验教学是高等药学院校最基本的教学形式之一，对培养学生科学的思维与方法、创新意识与能力，全面推进素质教育有着重要的作用。飞速发展的科学技术，已成为主导社会进步的重要因素。高等药学院校必须不断更新教学内容，以学科发展的前沿知识充实实验课程内容。

近年来，中国药科大学坚持以研究促教改，通过承担教育部"世行贷款——21世纪初高等教育教学改革项目"及立项校内教改课题等多种方式，调动了广大教师投身教学改革的积极性，将转变教师的教育思想观念与教学内容、教学方法的改革紧密结合起来，取得了实效。此次推出的国家"十一五"规划教材——药学专业双语实验教学系列，是广大教师长期钻研实验课程教学体系，改革教学内容，实现教育创新的重要成果。他们站在21世纪教育、科技和社会发展趋势的高度，对药学专业实验课程的教学内容进行了"精选"、"整合"和"创新"，强调对学生的动手能力、创新思维、科学素养等综合素质的全面培养。这套教材具有以下的特点：

1. 教材将各学科的实验内容进行了广泛的"精选"，既体现了高等药学教育"面向世界、面向未来、面向现代化"，也考虑到我国药学教育的现状与实际；既体现了各门实验课程自身的独立性、系统性和科学性，又充分考虑到各门实验课程之间的联系与衔接，有助于学生在教学大纲规定的实验教学学时内掌握基本操作技术，提高动手能力，养成严谨、求实、创新的科学态度。

2. 教材中新增的综合性、设计性实验有利于学生全面了解和综合掌握本门实验课程的教学内容。这一举措既满足了学生个性发展的需要，更注重培养学生分析问题、解决问题的能力和创新意识。

3. 教材中适当安排一些反映药学学科发展前沿的实验，有利于学生在掌握实验基本技术的同时，对药学学科的新进展、新技术有所了解，激发他们学习药学知识与相关学科的兴趣。

4. 教材以实践教学为突破口，采用双语体系编写，为实验课程改革构建数字化、信息化和外语教学的平台，有利于提高学生的科技英语水平。通过我校多年的药学系列实验课程双语教学实践，证明学生完全能够接受此套教材的教学。

国家十一五规划教材——药学专业双语实验教学系列教材的陆续出版，必将对推动我国高等药学教育的健康发展，产生积极而深远的影响。由于采用双语体系编写药学教学实验丛书尚属首次，缺乏经验，在内容选择及编写方法上的不妥之处，在所难免。欢迎从事药学教育的同行们批评赐教。

吴晓明

（中国药科大学校长、博士、教授、博士生导师）

Preface

Experimental teaching is one of the most fundamental teaching means in pharmaceutical colleges, playing an important role in training scientific thoughts and methods, creative consciousness and ability of the students as well as in promoting quality-oriented education in all-round way. Fast-advancing science and technology has come to be an important factor in dominating social progress. Teaching materials must be updated continually in pharmaceutical colleges, especially enriching the materials of experimental courses with the most advanced knowledge in the subject.

In recent years, China Pharmaceutical University have been stressing the promotion of teaching reform on the basis of research, succeeding in stimulating teachers' enthusiasm for teaching reform by various means such as undertaking the project of teaching reform in higher education at the beginning of 21st century sponsored financially by World Bank and entrusted by the Ministry of Education as well as approving and ratifying internal programs on teaching reform. Meanwhile, it yields fruits to integrate the transforming of teachers' educational ideology into the reform of teaching materials and methods. This series of textbook of national "11th five" planning-bilingual pharmaceutical experimental teaching series, is an important achievement made through studying ueaching system of experimental courses for long, reforming teaching materials and carrying out educational innovation of all the teachers concerned.

Meeting the new demands for education, science and technology and social growth, they select, integrate and innovate the teaching materials of pharmaceutical experimental courses, stressing the overall cultivation of comprehensive qualities, including experimental ability, creative thought and scientific attainments. This set of textbook possesses the following features:

1. These textbooks make an extensive "selection" of the experimental materials of each subject, reflecting the goal of facing the world, facing the future and facing the modernization in higher pharmaceutical education, and taking into account the status quota and reality of our pharmaceutical education; meanwhile embodying the individuality, systematicness and scientificalness of each experimental courses, which helps the students to grasp basic techniques of operation within the class hours of experimental teaching pre-

scribed by teaching syllabus and to improve their experimental ability and finally to cultivate a scientific approach of precision, practicality and creation.

2. The comprehensive designing experiments newly supplemented in the textbooks help the students to learn totally and grasp comprehensively the teaching materials of the experimental courses, which not only meets the students' needs for individual development but also trains their ability to analyze and solve problems and cultivates their creative consciousness.

3. Some experiments representing the latest development in pharmacy are properly included in the textbooks, which helps the students to learn about new advance and technology in pharmacy and to further arouse their interests in studying pharmacy and relevant subjects while grasping some basic techniques of experiment.

4. The textbooks take experimental teaching as starting point and are compiled in a system of bilingualism and aim to set up a platform of digitalization, information and foreign language teaching for the purpose of reforming experimental courses, which serves to enhance the students' level of technological English. It has been proved that the students have no difficulty being adapted to the teaching of this set of textbook through many years of bilingual teaching practice carried out in a series of pharmaceutical experimental courses of our university.

The successive publishing of the series of textbooks used for bilingual pharmaceutical experimental teaching-the national "11th-five" planning textbooks, will surely produce good and far-reaching influence in promoting the sound development of higher pharmaceutical education of our country. Since it is the first time that we have compiled this series of textbook of pharmaceutical teaching experiment in a bilingual system, we lack experience and thus some defects in choice of materials and way of compilation are inevitable. Experts engaged in pharmaceutical education are welcome to give any criticisms and advice.

Wu Xiaoming

Ph. D, prof., and supervisor of doctoral candidates
President of China Pharmaceutical University

前　言

　　电工电子学是一门非常注重实验的课程。自 20 世纪 80 年代初起中国药科大学就开始了本课程的教学实践，至今已经有近 30 年的历史。本课程作为药学类工科课程建设的一个重要部分，期间一直受到学校教务部门和有关院系的重视，也凝聚了许多老师的辛勤劳动。值得注意的是，这些年以来本课程理论内容和实验手段的持续不断更新，以及本课程的教学环境、教学对象、后续课程的要求改变等，构成了本课程的教学特点：课程的逻辑体系和知识体系需要并重，而且要简明扼要，或者说要注重教学效率。

　　本实验教材所涉及的实验平台、实验器材和实验电路都是经过精心选择和反复实践过的，也是药学、化学、生命科学类专业工科大学生必须掌握的基本知识和基本技能。因此编写者认为本书对学生及实验指导老师都具有实际的应用价值。本书所有内容采用中英文对照形式进行编写。从中国药科大学双语实验教学的实践看来这样的举动对学生是有益的和可行的。

　　参加本书编写的有张琪、张爱平、张利、钟宁。本书的英文编写工作受到了陈曙老师的帮助和指导。编者非常感谢多年来众多的学生在使用本书过程中所提出的宝贵意见和建议。编者特别需要感谢的还有 顾文照 教授、方醉敏教授、李剑鸣教授，他们一直进行着电工电子学课程的教学工作，直至前几年退休后还非常关心电工电子学实验室和实验教材建设。对中国医药科技出版社的编辑和中国药科大学教材科多年来提供的服务和帮助也表示感谢。

　　恳请读者对本书的错误和不妥之处提出批评指正。

张　琪

2007 年 3 月

PREFACE

Electrical engineering and electronics is a practical course of which experiment takes an important role. From the early 1980's on, China Pharmaceutical University (CPU) has begun the teaching practice of the course of electrical engineering and electronics. This course, as an important one among the courses of pharmaceutical engineering, has been taken seriously by the Educational Administration Bureau and other colleges of CPU, and also contained much of hardworking of teachers from CPU. It should pay more attention that the course contents and experiment means have been continually altered and innovated. More practically and specially, because of the teaching environment, students interesting and successor courses in the college like CPU, this course has its own characteristics: pay equal attention to logic system and knowledge system as well as be concise, or being taught efficiently.

All the lab platform, experiment equipments and circuits appearing in this book are been carefully selected and practiced for many times, and also are basic electrical knowledge and skills for engineering students major in pharmacy, chemistry and life science. Thus, we think this book has practical value for students as well as instructors in their lab workings. In this edition, we translate all the contents into English language. According to the bilingual teaching practice in CPU, it is profitable and feasible for college students.

The writers of this book are Qi Zhang, Aiping Zhang, Li Zhang and Ning Zhong. Mr. Shu Chen has given a lot of advices for the translation work of this book. We are delight to acknowledge students who used the previous editions for their practical and creative suggestions. We wish to express our appreciation to Prof. Wenzhao Gu, Prof. Zuimin Fang and Prof. Jianming Li, who have done so much teaching and constructional work for this course for many years, and more, after their retirement they are concerning the development of this course. We are also grateful to the Education Administration Bureau of CPU and National Medicine Press for their publishing service and other helpful work.

Any further suggestion is gratefully appreciated.

Qi Zhang
Mar 2007

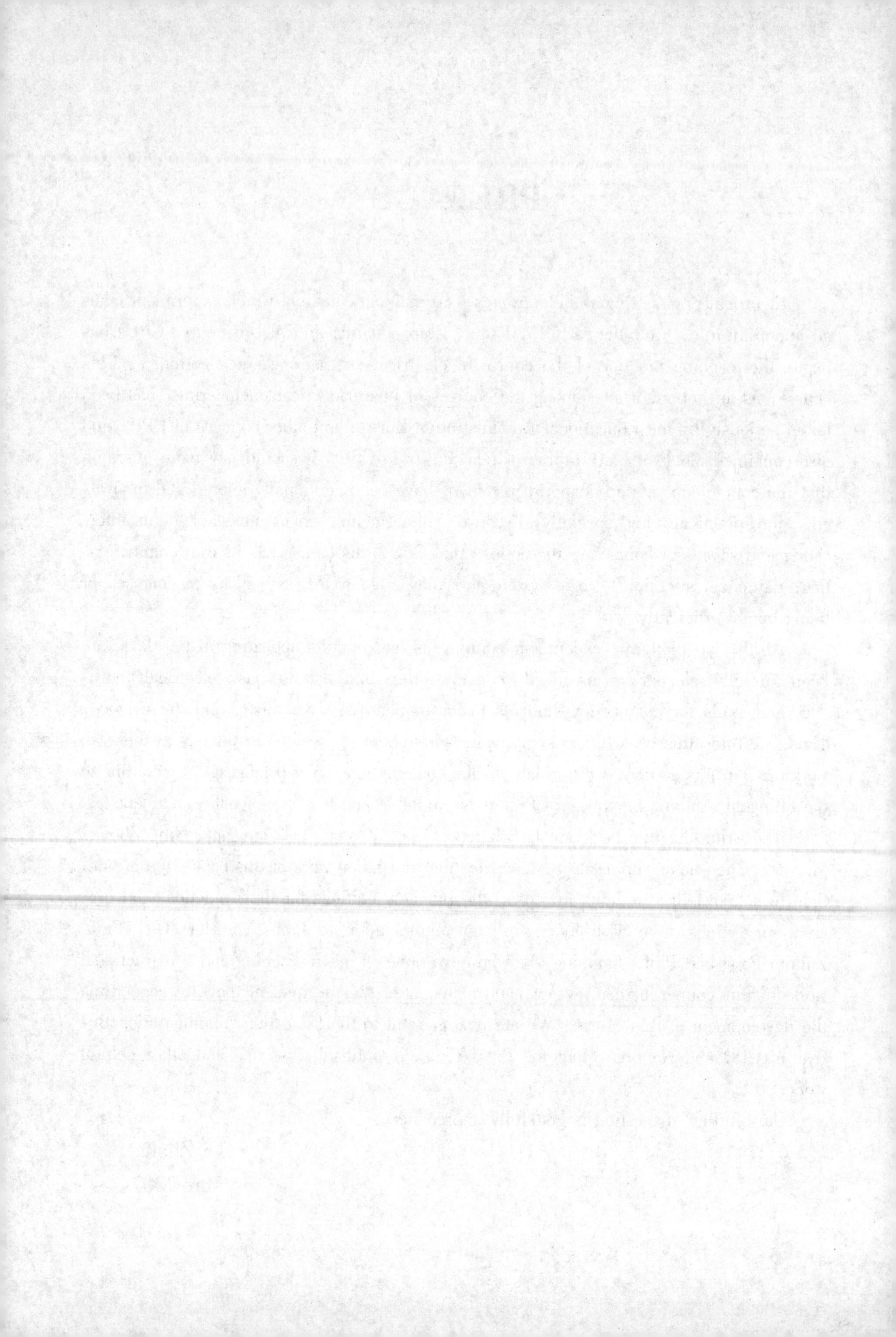

目　录
CONTENT

实验一　认识电路

【实验目的】

1．熟悉实验室用电规则和安全操作知识。
2．学会验电笔的使用。
3．利用一些常用电器件，练习电路连接的基本技能。

【实验原理与方法】

1．实验电路

本实验模拟一家庭用户电路布线情况，见图 1－1。

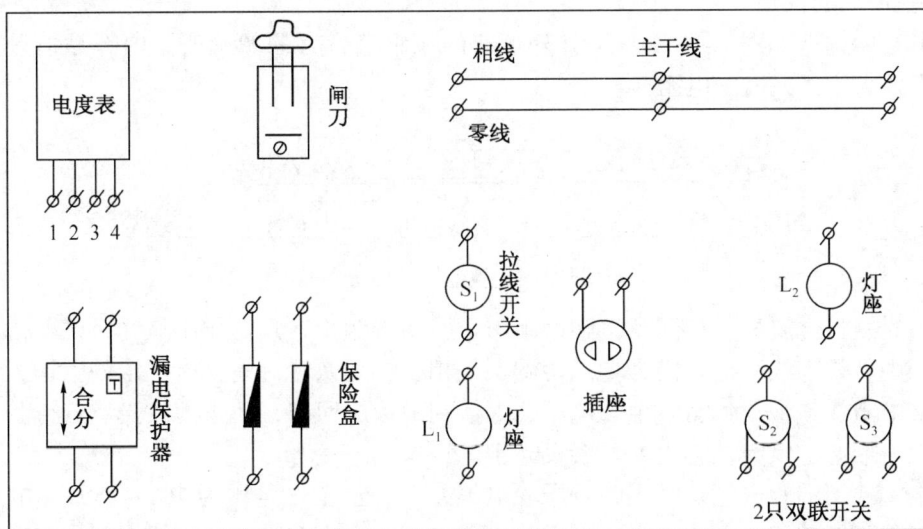

图 1－1　实验电路板

电源由总线进来，先经电度表、漏电保护器、闸刀、熔丝、主干线，然后分成三路。第Ⅰ路接单联开关控制的电灯，第Ⅱ路是插座，第Ⅲ路由两只双联开关控制一盏电灯。

（1）单相电源由一根相线和一根零线组成，按规定：相线接右边，零线接左边。实验中为方便检查，相线用红色导线，零线用蓝色或黄色导线，以示区分。

（2）插座：右眼应接相线，左眼接零线。

（3）漏电保护器，防止人身触电或回路泄漏的保护电器。当发生人身触电或回路有泄漏电流时，它能在 0.1 秒钟内将电源自动切断，保护人身安全，防止漏电造成事故。

另外，漏电保护器具有过电压保护功能，当电源电压突然升高到可能损坏用电器具时，它能迅速将电源自动切断。

（4）闸刀的作用是对电路总的控制，拉开闸刀，整个电路就没有电了，一般在对电路进行检修或改装时，应先把闸刀拉开。

（5）熔丝起短路保护作用，电路中发生短路故障时，电流很大，将熔丝烧断，电路就从电源断开，不致造成更大的危害。

（6）电度表用来对电能进行计量，表内两只线圈，电流线圈 I 与负载串联，电压线圈 V 跨接在电源上，即和电源并联，电表中转盘所指的圈数正比于用电的度数，圈数用计数器显示，每转若干圈代表 1 度，即 1kW·h。电度表的接线见图 1-2。

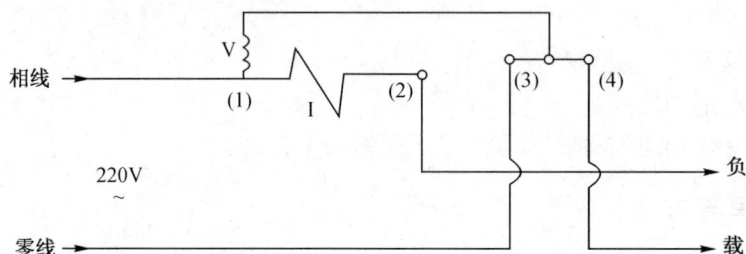

图 1-2　电度表接线图（1，3 进；2，4 出）

2．验电笔的使用

验电笔的构造如图 1-3 所示，其外形有的像钢笔，有的像旋凿。内部是一只氖泡串联一个阻值大于 1 兆欧的电阻。

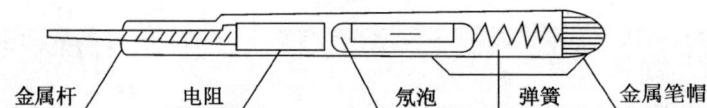

图 1-3　验电笔的构造

使用时，将金属杆与待测点接触，手与金属帽接触。若氖灯放出红光，说明待测点是相线，否则就是零线。这是因为如果待测点是相线，那么它对地就有一定的电位，电流经金属杆、电阻、氖泡、弹簧、金属笔帽、人体到地构成回路，使氖泡发光。若待测点为零线，则无电流通过氖泡，它也就不会发光了。

本次实验用的验电笔是常用的低压验电笔，其正常工作电压为 100～550V，电压超过这个范围就不能使用了。此外，在使用验电笔以前，应先在已知的相线上测试，确定氖灯能正常发光后再使用。验电笔使用时的正常握法如图 1-4 所示，应使笔尾金属体触及皮肤，但手指不可触及笔尖或旋凿金属杆部分，否则有触电危险。另外应使氖管小窗背光而朝向自己。

3．熔丝的选择和安装

（1）熔丝的选择

熔丝粗细是根据该线路上最大工作电流来决定的。选熔丝的额定电流略大于线路最大工作电流。

设本次实验中线路最大工作电流为 3 安培，则应选直径为多少毫米粗的铅锡合金熔丝？（查附录一《常用低压熔丝规格表》）

（2）熔丝的装接

装熔丝时必须先切断电源（因为闸刀开关的上桩头上有电），然后再动手装接。装接

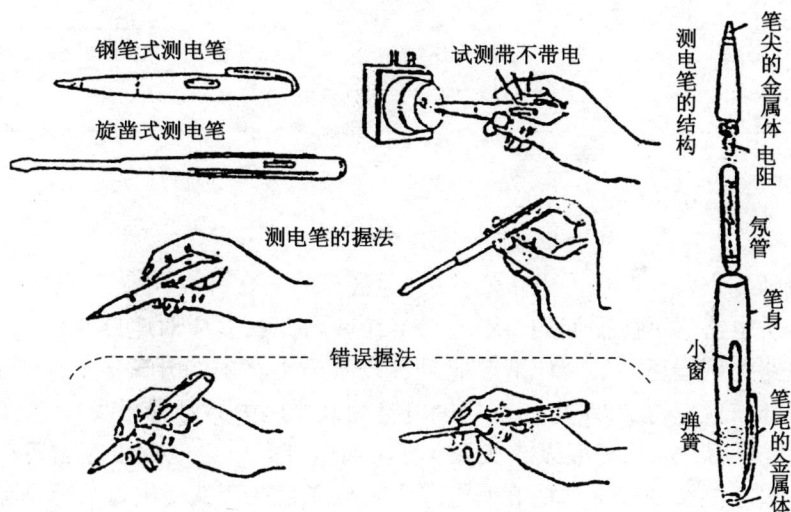

钢笔式测电笔
旋凿式测电笔
试测带不带电
测电笔的结构
笔尖的金属体 电阻
氖管
笔身
小窗
弹簧
笔尾的金属体
测电笔的握法
错误握法

图 1 – 4

时，保险丝按 S 形安装。如图 1 – 5 所示，拧螺丝时用力要适当，不能把保险丝压断。

【实验内容和步骤】

在安装前，先用验电笔测定单相交流 220V 电源，记住哪一根是相线，哪一根是零线，然后切断电源。在电工实验板上进行电路安装，经教师检查无误后方可合上电源。

1．电路连接

第Ⅰ路为电灯负载，因为要控制电灯的亮灭，所以要串接进开关 S_1。开关 S_1 要装到相线上（为什么？）

第Ⅱ路插座是用来向一些电器设备（如收音机、电视机、台灯等）供电。这些电器本身一般带有开关。所以这一电路中不再接开关。

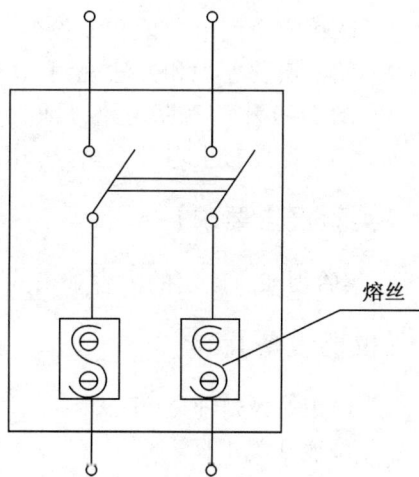

熔丝

图 1 – 5

第Ⅲ路，是用两只双联开关 S_2、S_3 控制一盏灯。双联开关与一般开关不同之处在于：它切断一条电路的同时，又接通另一条电路。两只开关分别安装在不同的地方，因此可以在两处对电灯进行控制，譬如楼梯灯，我们要求在楼上、楼下都能对灯进行控制。这一电路就可以做到：灯亮时，楼上楼下都能关掉它；而灯灭时，两处都可将它开亮。如图 1 – 6 所示。

在本实验中，主要是练习线路连线，所以就不按实际布线方式来连接电路，只要用导线将这些电路正确连接便可。

2．电路检查

电路连接好后，就要检查连接是否正确，一般要按电路图复查一遍线路，看看有什么地方接错了没有。这一步检查手续，对我们初学电工的同学来说是必须的。在自检结束

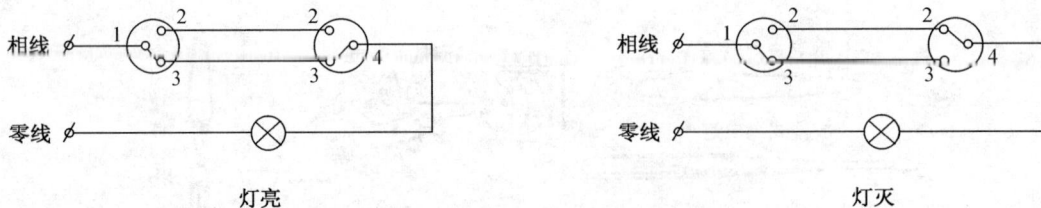

图 1-6 两只双联开头控制一盏灯电路图

后，还要经教师复查。在以后的电工实验中，我们都要这么做。电路连接检查无误后，就可通上电源，看电路能否按预期的要求进行工作。因此我们应对电路正常工作心中有数。以本实验的电路为例，正常的工作情况应是：控制第Ⅰ路中的开关 S_1，可以使电灯 L_1 亮灭。第Ⅱ路中插座上应有 220V 电压（用验电笔测试插座中哪侧为相线）。在第Ⅲ路中，不论 S_2 或 S_3，都能使灯 L_2 从亮变灭或从灭变亮。而电灯亮时，电度表应当转动，对用电量进行计量。如果不能实现以上功能表明电路中有故障存在，就要进行故障排除。

【实验预习要求】

1. 阅读安全用电常识和实验须知。
2. 阅读实验原理和方法说明。
3. 图 1-1 中，三路电路（即三个负载）是并联还是串联？
4. 如果电路接上电源，熔丝烧断，应如何检查和处理？

【实验报告要求】

按实验须知所确定的内容和格式填写实验报告，并回答预习要求中提出的问题。

【仪器设备】

1. 白炽灯及灯座 2 套。
2. 拉线开关 1 只。
3. 漏电保护器 1 只。
4. 闸刀开关 1 只。
5. 交流电压表 1 只。
6. 熔断器 2 只。
7. 双联开关 2 只。
8. 熔丝 1 根。
9. 电度表 1 只。
10. 验电笔 1 支。

1 Basic Knowledge about Electric Circuits

Objectives

1. Be familiar with electric lab rules and electric safeguard.
2. Learn to use the electric detector pen rightly.
3. Learn to join some circuits with the common electric devices.

Principle and Method

1. The experiment circuit

The circuit in this experiment simulates the wiring of a family, as shown in Figure 1 – 1.

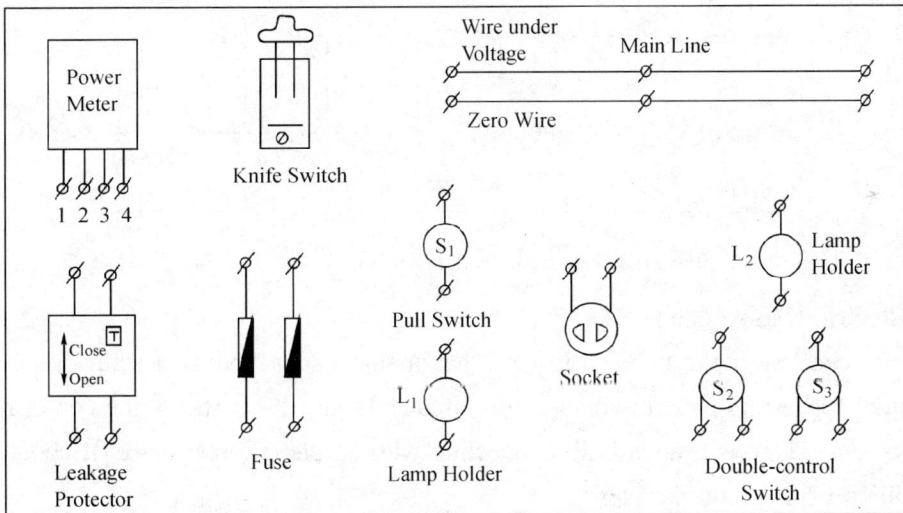

Figure 1 – 1   A lab electric circuit board

The electric source line goes through kilowatt-hour meter, the leakage protector, knife-switch, fuse and trunk line, then falls into 3 loop: the first connects with a lamp which is controlled by a switch; the second connects with a socket; and the third connects with a lamp which is controlled by a double-control switch.

(1) The single-phase electric source is made up of a wire under voltage and a zero wire, and it is a regulation that the wire under voltage should be set on the right hand and the zero on the left hand. Practically, for the sake of distinguishing the wire under voltage should be joined by red wire and the zero wire by blue or yellow.

(2) The socket: the holes on the right connect with wire under voltage and the holes on the left connect with the zero wire.

(3) Leakage protector. A leakage protector is a safeguard for electric shock or leak of electric current. When electric shock or leak of electric current in a loop happens, the protector can cut off the electric source in 0.1 second automatically.

In addition, a leakage protector has a function of keeping electric appliance out of damage when the source voltage increases suddenly by cutting off the source at once.

(4) The knife-switch can control the whole circuit. Withdraw the knife-switch, the whole circuit becomes out of energy. When exam or repair the circuit, the knife-switch should be pull off.

(5) The function of a fuse is when there is a fault such as short in the circuit the current increases and the fuse will be melt to cut the source, avoiding more damage.

(6) The kilowatt-hour meter is a device to measure the electric energy consumption in the circuit. Figure 1 − 2 shows its connection way in a circuit.

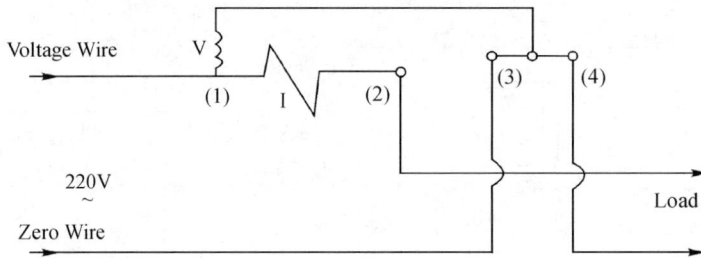

Figure 1 − 2   The wiring diagram of the kilowatt-hour meter (1, 3 in, 2, 4 out)

2. Electric testing pen

Electric checking pen, or electric pen, has a shape of a pen or a screwdriver; it is a common tool for testing electric voltage in a circuit. Figure 1 − 3 shows the conformation of an electric pen. There is a neon bulb connecting with an electric resistance of about not less than 1 million Ohm inside the pen.

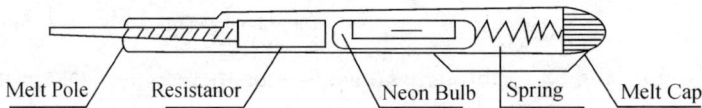

Melt Pole   Resistanor   Neon Bulb   Spring   Melt Cap

Figure 1 − 3   The conformation of an electric pen

To check a certain point in a circuit, put the metal pole in contact with the point while your naked hand touching with the metal cap in the other end of the pen. If the neon bulb shins with red light, the testing point is a wire under voltage; or, it is a zero line. Because contacting with a point of under voltage a very small electric current can flow through the loop of metal pole, the resistance, the neon bulb, spring, metal cap and your body to make the bulb shine.

The electric pen in our lab course is a low voltage pen of which the working range is 100 ~ 550V, so this kind of the electric pen cannot be used under situation beyond the range. It is suggested that the pen should be checked on a wire under voltage before it is put into work. Figure 1 – 4 shows the correct manner of holding the pen, where your finger should contact with the metal part at the end of the pen but be sure not contact with the metal pole, or it may be under the dangerous of electric shock. Let the neon bulb window expose to yourself while checking with the pen.

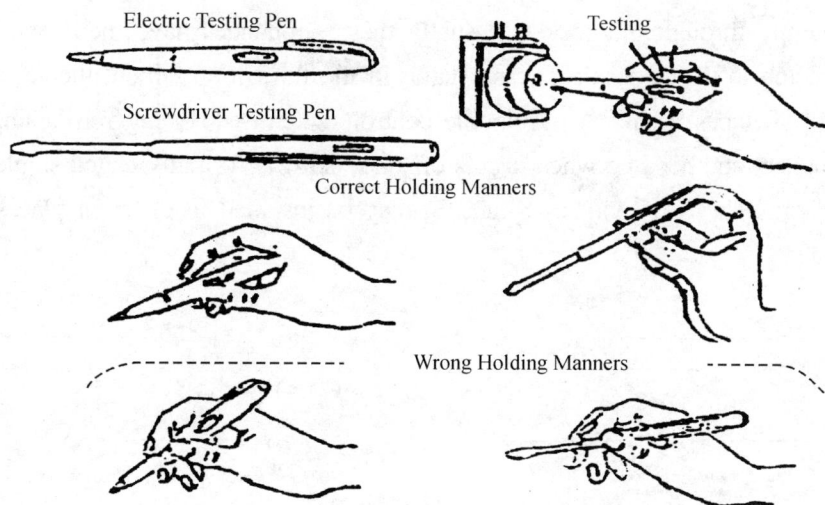

Electric Testing Pen

Testing

Screwdriver Testing Pen

Correct Holding Manners

Wrong Holding Manners

Figure 1 – 4

3. Fuse

(1) Choice of the fuse

The diameter or size of a fuse is determined by the maximum electric current in a certain circuit. The rated current of a certain fuse must be a bit great than the maximum current of the circuit.

The maximum working current in this experiment is 3A, and please consult the table "The specification of low voltage fuse" in Appendix 1 to make a suitable choice.

(2) Fixing of a fuse

Before install a new fuse please make sure to cut off the electric current. The fuse wire should be bended in the shape of "S", as shown in Figure

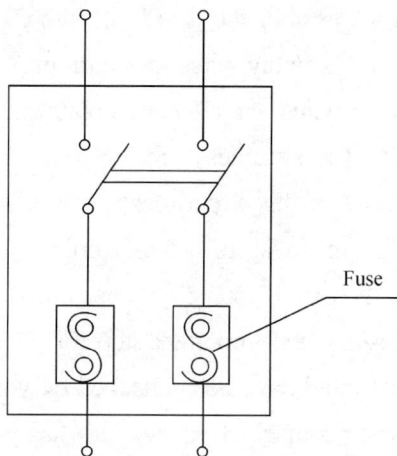

Fuse

Figure 1 – 5

1 – 5. Properly drive the screw or the fuse may be broken.

Experiment

The electric source is single phase AC 220V, and check the wire under voltage by your electric testing pen first. Be aware that the source is cut off while you are joining the circuit. All the switches must be set on the wire under voltage. (Why?) The knife switch can only be put on after your instructor has checked your circuit and agree to do so.

1. The first loop has a load of lamp, which is controlled by the switch S_1 connected in series. The purpose of the second loop is power supply to some appliance such as radio, TV set or desk lamp, through the socket. Usually these appliances have their own switch, so there is no switch in this loop. There is a lamp in the third loop which is controlled by the double-control switches S_2 and S_3. A double-control switch is different from a single switch, which can connect another line when it cuts off one, so it is formally called single-pole double throw switch. The two switches S_2 and S_3 may be installed in different places to control one appliance.

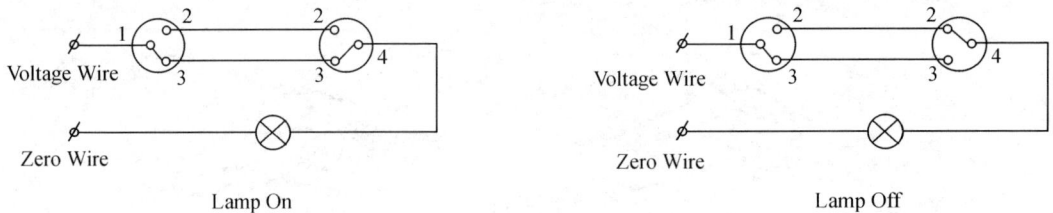

Figure 1 – 6　One lamp is controlled by a double-control switch

Actually the knife-switch, fuse and kilowatt-hour meter are installed together on one board near the entrance of power supply bus, and the board is called electric supply board. The wires after the board fall into different loops connecting with certain appliance. There are two ways to lay wires in a building: hidden tub laying and open laying. With hidden tub laying the wires are all put into tubes. And with open wire laying the wires are fixed on the wall. The switches, lamp holders and sockets are set in places where they are needed. The purpose of this experiment is to practice joining correct circuit, so it is unnecessary to join a circuit in a way of actually doing.

2. After your circuit is joined, it should be checked with the circuit diagram. It is a necessary and important step for a learner of electric engineering. After self-check, your circuit should have been checked by your instructor. This checking process cannot be omitted in this lab course. If your circuit has passed all the checks you can put on power to see whether the circuit works properly as you expecting. Therefore you should know fairly well about the working states of your circuit. The normal working states of the circuit in this experiment

should be: the switch S_1 can control the lamp L_1 in the first loop; the socket in the second loop has a voltage of 220V and the wire under voltage is on the right side; in the third loop the switch S_2 or S_3 can put the lamp L_2 into bright or dark; and the kilowatt-hour meter should be running when there is a bright lamp. If there is any fault in your circuit, you should have to exam the circuit again to exclude the fault and put it working properly.

Preparation requirement

1. Reading electric safety knowledge and experiment notice.
2. Understand the principle and working instruction of this experiment.
3. In Figure 1 – 1, how about the three loops, in series or in parallel?
4. If the fuse is melt broken, how to exam and deal with?

Report requirement

Write your experiment report in the right format and content as it is confirmed in the preface of the text. Answer the questions in the preparation requirement.

Equipments

1. Filament lamp and holder, 2 sets.
2. Cord-pull switch.
3. Leakage protector.
4. Knife-switch.
5. AC voltmeter.
6. Fuse box, 2 sets.
7. Single-pole double control switch, 2 sets.
8. Fuse.
9. Kilowatt-hour meter.
10. Electric testing pen.

实验二　日光灯电路和功率因数的提高

【实验目的】

1. 掌握提高电路功率因数的方法，理解提高功率因数的意义。
2. 了解日光灯工作原理，学会日光灯电路连接。
3. 学习功率表的使用。

【实验原理与方法】

1. 交流电路的功率因数

在交流电路中由于存在着电感性负载和电容性负载，电路中瞬时电压与电流相位通常不一致，因而存在视在功率 S、有功功率 P 和无功功率 Q 三种不同的功率。

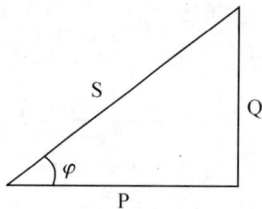

$$S = UI$$
$$P = UI\cos\varphi$$
$$Q = UI\sin\varphi$$

交流电路中功率因数的大小为 $\cos\varphi = P/S$，它关系到电源设备及输电线路能否得到充分利用。$P = UI\cos\varphi$，从供电方面来看，当线路的功率因数较低时，在同一电压下输送给负载一定大小的有功功率时，所需的电流就较大，增加了线路功率损耗；另一方面功率因数愈低，发电机所发出的有功功率就愈小，而无功功率却愈人，发电设备的容量愈得不到充分利用。因此提高线路的功率因数既可提高电源设备的利用率，又可减少线路的能量损失。

本实验在日光灯电路中由于镇流器是电感性负载，电路的功率因数较低，为了提高图 2-1 电路的功率因数，可在 a、b 两端并联一电容，见图 2-2。当并联电容后，对于原感性负载来说（图 2-2 虚线框内），所加电压和负载参数均未改变，即没有改变电路的工作

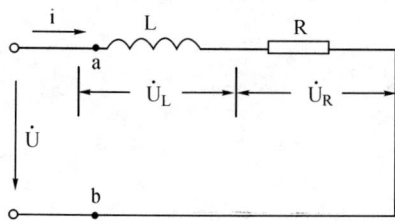

图 2-1

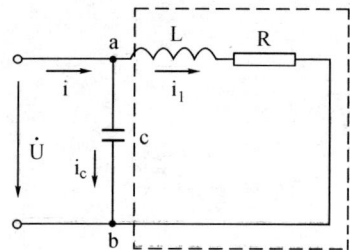

图 2-2

状况。但是并联电容后，由于 I_C 的出现，电路的总电流（即电源向外输送的电流）减小了，见图 2-3。即线路的功率因数得到了改善（φ 角减小，$\cos\varphi$ 增大）。

由上述分析可知，并联电容前、后，①电源向外供出的有功功率没变（负载的有功功率没变），②总电流因并联电容而减少。

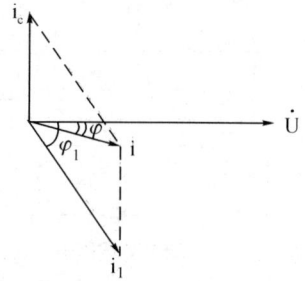

图 2-3

2．日光灯电路

日光灯电路由灯管、镇流器、启辉器组成。见图 2-4。

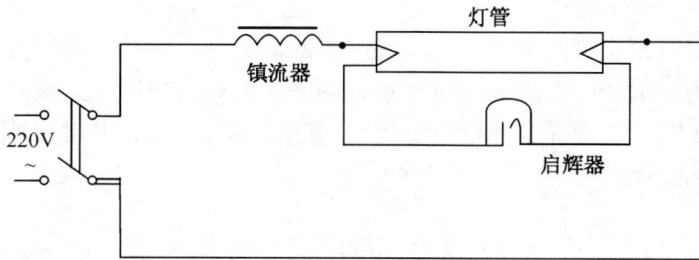

图 2-4

①灯管：在玻璃管内壁涂以荧光粉，管内充以氩气和少量的汞。灯管两端有灯丝。它须有一瞬时高电压帮助起燃。在正常工作时，灯管两端电压比较低，需要有一降压元件（镇流器）与它串联才能接于 220V 电源上正常工作。在实验电路中将灯管当作一电阻性元件。

②镇流器：为一有铁芯的电感线圈。其在电路中有两个作用：一是在灯管起燃瞬间产生一高电压帮助灯管起燃；二是在正常工作时，其上降去一部分电压，以限制通过灯管的电流，它在电路中属于有电阻的感抗元件。

③启辉器：启辉器在电路中起一自动开关的作用。在充有氖气的玻璃泡内装有两个电极。一为固定电极，一为用双金属制成"η"形的可动电极。平时动静触头不接触，当两电极加以一定高的电压时，则氖气电离形成气体导电。同时伴有热量产生使双金属片受热膨胀而与固定电极接触。一旦电极接触，气体导电就停止，双金属片不再受热而收缩恢复原来的断开状态。

日光灯工作过程如下：图 2-5，当闭合电源开关，此时由于日光灯管未起燃而不能导电。电源电压通过镇流器、灯管的灯丝施加于启辉器两电极上。启辉器两极间气体导电，双金属片与固定电极接触。当两电极接触不再产生热量，双金属片冷却复原，使电路突然断开。由于电路电流突然消失，镇流器产生一较高的自感电势经回路施加于灯管两端，驱使灯管起燃。此时电流经镇流器、灯管而流通。由于灯管起燃后，其两端压降较低，约 40～100V 左右，启辉器不再动作，日光灯进入正常工作状态。

3．功率表的使用

功率表是一种电动式仪表，可用于测量直流电路和交流电路的功率。它有两组线圈，一组是电流线圈，匝数较少，导线较粗，串接于被测电路中；另一组是电压线圈，匝数较多，导线较细，和附加电阻相联后并接在电路中。瓦特表的指针偏转是与电压、电流以及

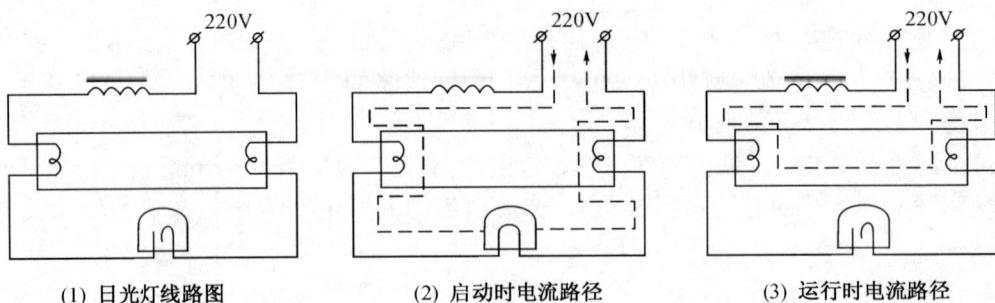

| (1) 日光灯线路图 | (2) 启动时电流路径 | (3) 运行时电流路径 |

图 2 - 5

电压、电流之间的相位差角的余弦成正比的，即和被测电路的有功功率成正比，因此可用它测量电路的功率。

①功率表使用时，应将两个"∗"端接向电源。若发现指针反向偏转，可使用表面专用旋钮将指针改为正偏转。电流线圈应和负载串联，电压线圈应和负载并联。

②选用功率表时，要选择合适电流量程 I_N 和电压量程 U_N，电流表高低量限连接及功率表连接如图 2 - 6 所示。

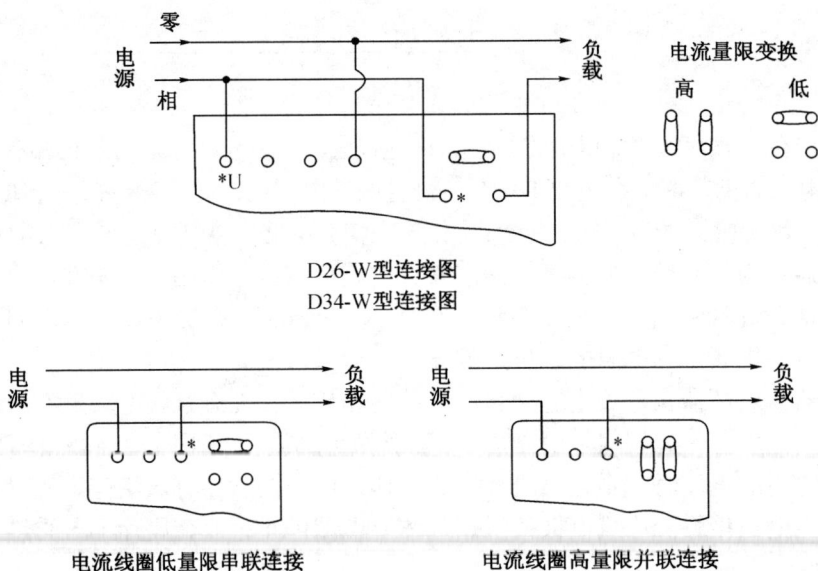

D26-W型连接图

D34-W型连接图

电流线圈低量限串联连接   电流线圈高量限并联连接

图 2 - 6

③注意不同类型功率表仪表分格常数 C 的查阅方法，见附录二《功率表的使用》。

【实验内容和步骤】

1. 测定电路参数

按图 2 - 7 接线，观察日光灯的启动情况，测定 U、U_L、U_R、I、I_1、P，记入表 2 - 1，并求出日光灯电路的功率因数 $\cos\varphi$：

$$\cos\varphi = \frac{P}{UI} \qquad 灯管等效电阻 \ R = \frac{U_R}{I_1}$$

镇流器参数：$r = P/I_1^2 - R$，$X_L = \sqrt{(U_L/I_1)^2 - r^2}$，$L = X_L/\omega$

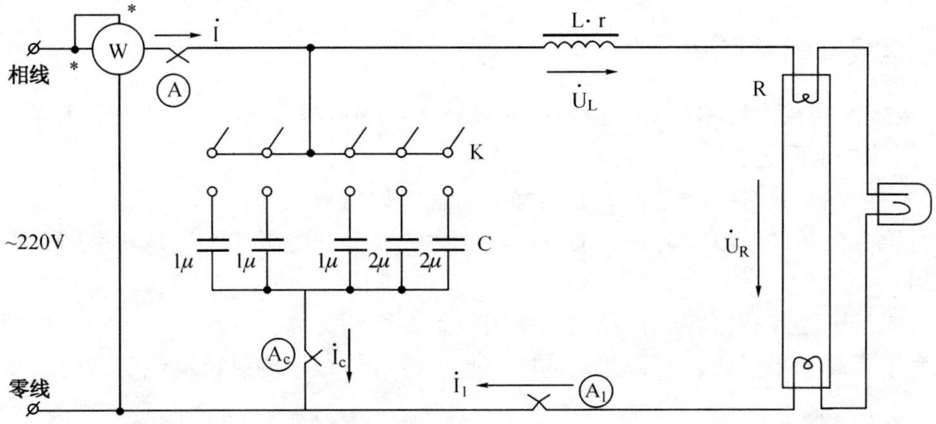

图 2-7

表 2-1

C (μF)	测量值							计算结果			
	U (V)	U_L (V)	U_R (V)	I (A)	I_1 (A)	I_C (A)	P (W)	$\cos\varphi$	R (Ω)	r (Ω)	L (H)
0											
1											
2											
3											
4											
5											
6											
7											

2．将电容 C 与日光灯支路并联

接通电源，将 C 从零逐渐增加到 $7\mu F$，使电路从感性变到容性。每改变 C 一次，测出 U、U_L、U_R、I、I_1、I_C、P，记入表 2-1。

3．在改变电容 C 的过程中，观察表 2-1 中的 U、U_L、U_R、I_1 和 P 有无变化，说明为什么？

4．注意事项

①日光灯的启动电流较大，应等待日光灯点燃后，再将电流表插入。

②测量电流的电插头其金属部分带电，操作中切不可触及金属棒部位。

【**实验预习要求**】

1．复习有关"功率因数的提高"内容，熟练掌握用相量图分析 R、L、C 的并联电路。

2．阅读附录部分《功率表的使用》。

3．了解日光灯工作原理，弄清日光灯点亮前和点亮后的电流通路。

4．掌握仪表的正确连接和测量方法，知道如何由功率表的读数求得实际的功率数。

【实验报告要求】

1．按规定格式内容编写实验报告，完成表 2-1 中 $\cos\varphi$、R、r、L 的计算值。

2．回答如下问题：

①你在连接日光灯线路过程中遇到哪些问题？如何处理的？

②对感性负载并联电容为什么能提高功率因数？并联电阻能否提高功率因数呢？为什么不采用？

③日光灯电路并联电容后自身的功率因数是否得到提高？

④为了提高功率因数，是否并联的电容越大越好？根据测量数据说明理由。

【实验仪器】

1．日光灯电路实验板 1 块。

2．交流电压表 1 只。

3．交流电流表 1 只。

4．低功率因数功率表 1 只。

2　Fluorescent Lamp Circuit and the Improving of Power Factor

Objectives

1. Studying the method for improving power factor, understanding the significance of power factor.

2. Learning the working principle of fluorescent lamp, correctly joining a circuit containing fluorescent lamp.

3. Learn the operation of a power meter.

Principle and Method

1. Power factor of AC circuit

In an AC circuit, because of the inductive load and capacitive load, there may be a phase difference between instantaneous voltage and instantaneous current. It is necessary to study the power in an AC circuit in three types: apparent power S, real power or active power P, reactive power Q.

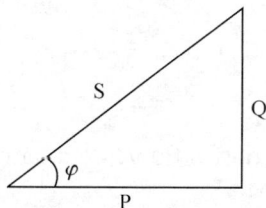

$$S = UI$$
$$P = UI\cos\varphi$$
$$Q = UI\sin\varphi$$

The power factor of AC circuit is defined as: $\cos\varphi = P/S$, it is an utilization coefficient about the source and transmission equipment. $P = UI\cos\varphi$, under the situation of low power factor, it may cause a great current to supply a certain active power to load under a certain voltage; or an electric generator can only supply a low factor of active power and the generator capability cannot be sufficiently put into work. Therefore improving the power factor can not only improve the utilization of power supply equipments but also reduce the energy loss in circuit.

In this experiment, because the rectifier in the fluorescent lamp circuit is an inductive load, the power factor is quite low. To improve the power, we can join a capacitor between the point a and b, see Figure $2-1$. After install the capacitor, the parameters about the voltage and load of the original inductive elements (marked in the dots line frame in Figure 2 – 2) have not been changed, thus the elements work under the same situation as before. But because of the present of the current I_C, the total current i (the current feeding by the power supply) becomes smaller, see Figure 2 – 3. That is the power factor has been improved (the angle φ becomes smaller and the $\cos\varphi$ larger).

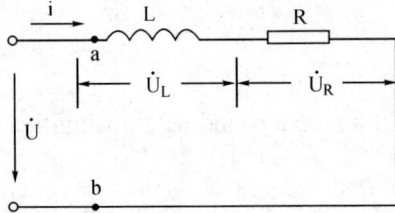

Figure 2 – 1

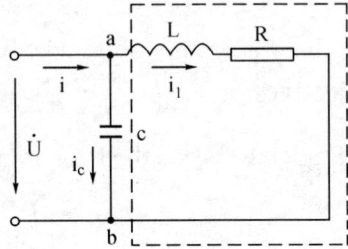

Figure 2 – 2

So, after installed a parallel capacitor in the circuit, we know: first, the actual power supplying by the source has not been changed; second, the total current reduced.

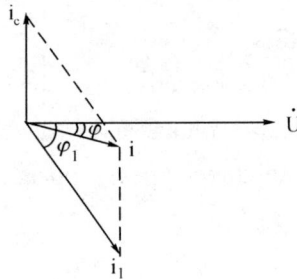

Figure 2 – 3

2. Fluorescent lamp circuit

In Figure 2 – 4, the fluorescent lamp circuit is consisted of a fluorescent lamp, a rectifier and a starter.

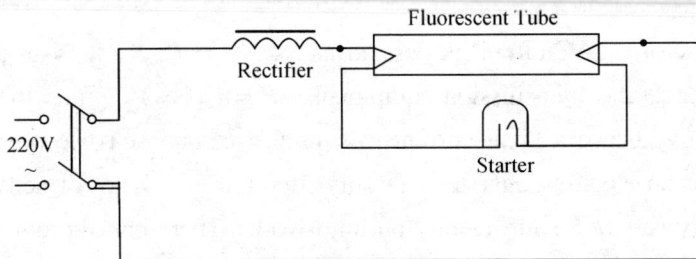

Figure 2 – 4

① The lamp tube. The glass tube is filled with argon and a small quantity of mercury and its inner surface is painted by fluorescence powder. There are filaments at the two ends of the tube. To brighten the tube, there must be a high voltage to blaze the gas. When working normally, there must keep a quite low voltage upon the two ends of the tube. So, there is a rectifier in series in the circuit to keep the tube works normally when the lamp is connecting to 220V electric source. In this experiment the tube can be treated as a resistance element.

② Rectifier. A rectifier is an inductance solenoid with a iron core in the center. It has two functions in the circuit: one is to produce a high voltage to brighten the tube; another is to reduce the voltage upon the tube to keep the lamp working normally. In this experiment, the rectifier is an inductive element in the circuit.

③ Starter. The starter in this circuit is something like an automatic switch. The starter is a glass bulb filling with neon gas with two metal poles. One pole is fixed and another in shape of "η", which is made up of two different metal flakes and is moveable. Normally the two poles cannot contact with each other. When the poles bearing a high voltage, the neon gas becomes ironed and conducts electricity, the moveable pole is heated and expands to contact with the fixed pole. As soon as the poles contact with each other, the gas stops conducting electricity and the double metal flake is not heated anymore. So the moveable pole will contract to break the electricity through the starter.

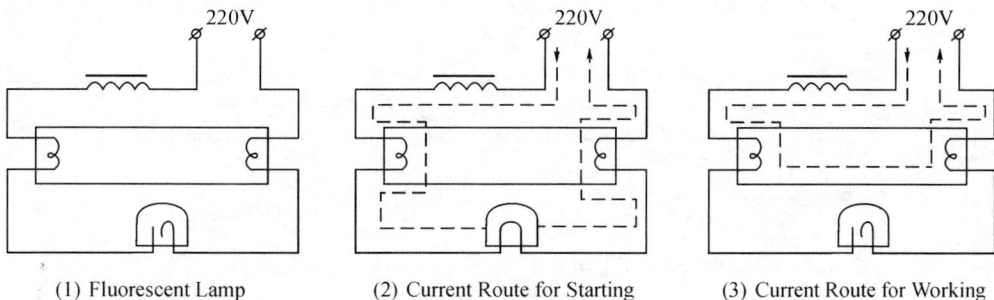

(1) Fluorescent Lamp   (2) Current Route for Starting   (3) Current Route for Working

Figure 2 – 5

The working process of fluorescent lamp is as following: when switch on the electric source the fluorescent tube is not bright because the gas in it is not blazed. Then the source voltage is falling on the starter through the rectifier and the filaments in the tube. The conducting-break action of the poles in the starter will cause a sufficient high self-inductive voltage upon the two ends of the rectifier falling on the filaments in the tube, which will cause the tube brightening. And now the current fellows through the loop of rectifier and filaments, the voltage between the two ends of the tube is about of 40 ~ 100V, which cannot cause action of the starter. The fluorescent lamp works in functional mode from now on.

3. Power meter

Power meter is a kind of electric meter and can measure the power consumption in a DC or AC circuit. It has two coils: one coil, which is called current coil, with less turns is joined in series in the circuit; another, which is called voltage coil, with rather more turns connecting with a resistance is joined in parallel in the circuit. The hand deflection of power meter is directly proportional with voltage, current and the cosine of the angle between voltage and current, or directly proportional with the actual power consumption in the circuit. So it is a suitable device for measuring actual power.

① When power meter is put into use, the two wiring posts marking by * should be joined with the wire under voltage. If the hand deflects reversely, one can turn the special knob on surface to make it turn correctly. The current coil should be in series with load, and the voltage coil should be in parallel with load.

② When measure power with a meter, it must pay attention to select suitable current rang I_N and voltage rang U_N. The connecting manner about small current rang and large current rang is shown in Figure 2 – 6.

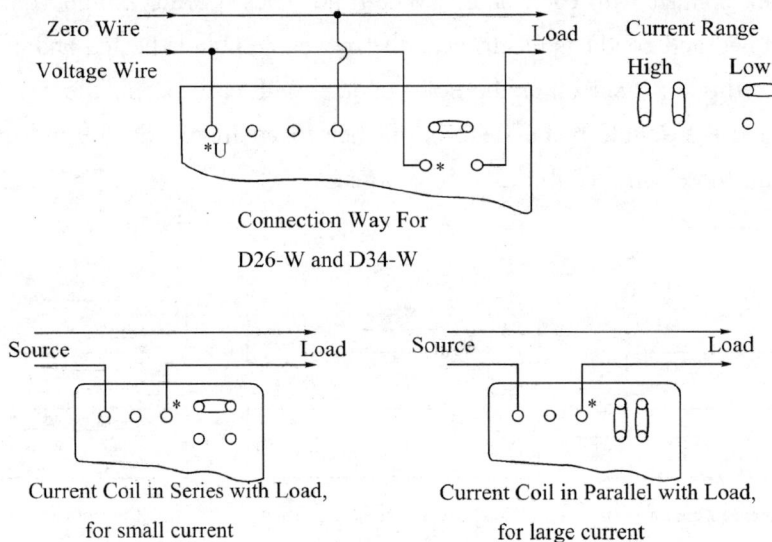

Connection Way For
D26-W and D34-W

Current Coil in Series with Load,
for small current

Current Coil in Parallel with Load,
for large current

Figure 2 – 6

③ Different type of power meter has different division constant C, one should consult beforehand. Appendix 2 has a detail about power meter.

Experiment

1. Measure the parameter of the circuit.

Join the circuit shown in Figure 2 – 7. Put on the source and observe the starting of the fluorescent lamp. Measure U, U_L, U_R, I, I_1 and P, record the values in Table 2 – 1, and calculate the power factor $\cos\varphi$.

$\cos\varphi = P/UI$,

Equivalent resistance of the tube: $R = U_R/I_1$

Parameter of the rectifier:

$$r = P/I_1^2 - R, \quad X_L = \sqrt{(U_L/I_1)^2 - r^2}, \quad L = X_L/\omega$$

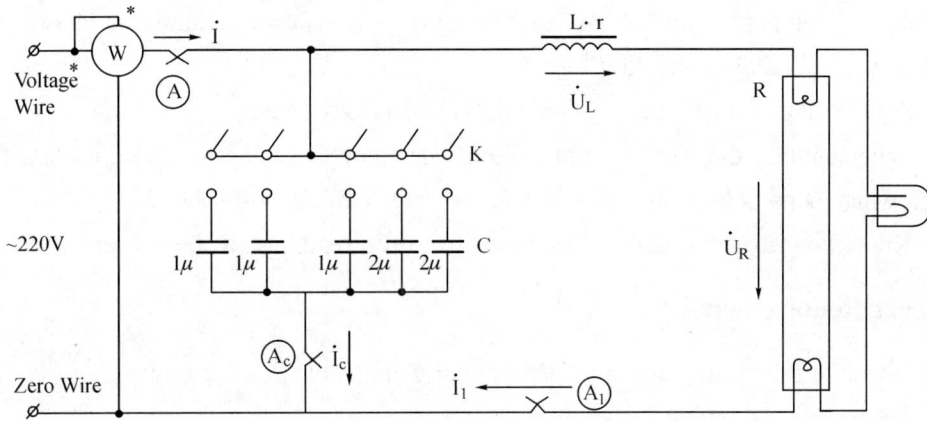

Figure 2 – 7

2. Set capacitor C in parallel with the fluorescent lamp

Change the capacitor C from zero to $7\mu F$ with one step of $1\mu F$, and then the circuit will change from inductive to capacitive. Record the corresponding values of U, U_L, U_R, I, I_1 and P in Table 2 – 1.

3. In the process of the alternation of the capacitor C, what happens to the values of U, U_L, U_R, I, I_1 and P?

Table 2 – 1

C (μF)	Measurement values							cosφ	Calculation results		
	U (V)	U_L (V)	U_R (V)	I (A)	I_1 (A)	I_C (A)	P (W)		R (Ω)	r (Ω)	L (H)
0											
1											
2											
3											
4											
5											
6											
7											

4. Notice

A) The brightening current of the fluorescent lamp is quite large, so the ammeter should have been plug into the circuit when the lamp is working normally.

B) The metal part of the plug connecting to the ammeter is electrified, be aware never touch it.

Preparation Requirement

1. Review the part of text about "improving of power factor", draw the phasor diagram of R, L and C parallel connection circuit.

2. Reading the Appendix 2: "Manipulation of Power Meter".

3. Understanding the working principle of the fluorescent lamp, make it clear that the current flowing loops before and after the fluorescent lamp is brightened.

4. Know how to get actual power value from the readings of the power meter.

Report Requirement

1. Calculate the values of $\cos\varphi$, R, r and L in Table 2 − 1.

2. Answer the following questions:

A) What problems have you met in this experiment? How do you deal with them?

B) Why the power factor can be improved by paralleling capacitor to inductive load? Why not resistance?

C) Have the power factor of fluorescent lamp loop itself been improved by paralleling capacitor?

D) In the interest of improving power factor, is it the larger of the capacitor the better? State your reason basing on your measurement.

Equipments

1. Fluorescent lamp circuit board.

2. AC voltmeter.

3. AC ammeter.

4. Power meter.

实验三　三相交流电路

【实验目的】

1. 学习负载星形和三角形的连接方法。
2. 验证三相电路中的线、相电量之间的关系。
3. 了解三相四线制的中线作用。

【实验原理与方法】

1. 三相四线制电源

对称的三相四线制电源的线电压 U_L 和相电压 U_P 之间数值上的关系为：

$$U_L = \sqrt{3}\, U_P$$

U_L：线电压，即任意两相线之间的电压值。

U_P：相电压，即任一相线与中线之间的电压值。

三相四线制的电压值一般是指线电压的有效值，如"三相 380V 电源"指线电压为 380V，其相电压为 220V。本实验用到"三相 220V 电源"是指线电压为 220V，其相电压则为 127V。

2. 负载作星形连接

（1）三相对称负载作星形连接，线电压 U_L 为相电压 U_P 的 $\sqrt{3}$ 倍。其线电流 I_L 等于相电流 I_P。

$$U_L = \sqrt{3}\, U_P$$

$$I_L = I_P$$

这种情况下，流过中线的电流 $I_0 = 0$，所以可以省去中线。

（2）三相负载不对称作星形连接时，必须采用三相四线制接法，以保证三相不对称负载的每相电压维持正常。倘若此时中线断开，三相负载各相电压不再对称，会引起负载不能正常工作。所以，在负载不对称时不可随意将中线断开。

3. 负载作三角形连接

负载作三角形连接时，不论负载对称与否，其相电压均等于线电压：

$$U_L = U_P$$

（1）三相负载对称作三角形连接时，其相电流也对称，相电流和线电流之间为：

$$I_L = \sqrt{3}\, I_P$$

（2）当三相负载不对称作三角形连接时，相、线电流不再是 $\sqrt{3}$ 倍的关系。

即

$$I_L \neq \sqrt{3}\, I_P$$

但只要电源的线电压 U_L 对称，加到三相不对称负载上的每相电压仍是对称的，各相负载均能正常工作。

【实验内容和步骤】

1．测量三相四线制电源的线、相电压值

对本实验所需的"三相220V"电源的相、线电压数值进行测量，并记入表3－1中。

因本实验中所用的三相负载为白炽灯，灯泡额定电压为220V，为了使三角形接法时电源电压不致过高，故三相电源380V的线电压经三相调压器降低到220V。即调出 3×220V 电源供本实验的使用。

表 3－1

电　压	U_{AB}	U_{BC}	U_{CA}	U_{AN}	U_{BN}	U_{CN}
220V 三相电源						

三相调压器接法如图 3－1

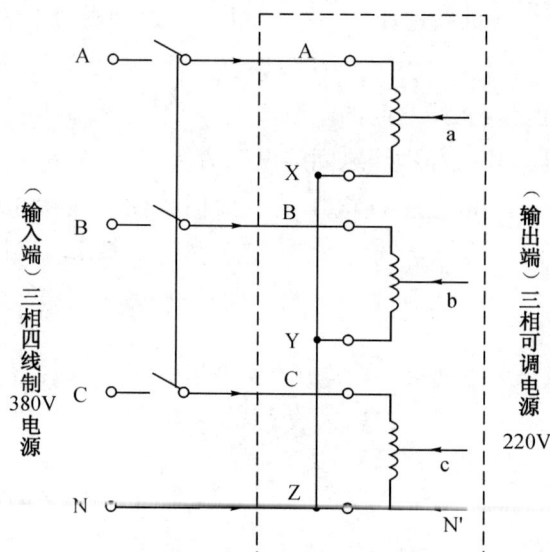

图 3－1

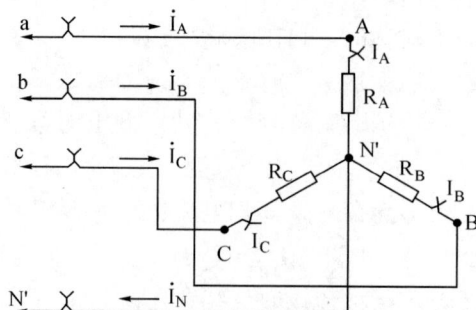

图 3－2

2．三相对称负载星形连接

用三只 40W 白炽灯泡按图 3－2 接成星形三相对称负载，然后按有中线和无中线两种情况进行实验。电源电压调在"三相 220V"。

（1）有中线

线路接好后，经教师检查无误后合上三相电源闸刀，用电压表和电流表测量各电压和电流值，记入表 3－2。

表 3－2　负载 Y 形连接

		Y 负载对称		Y 负载不对称	
		有中线	无中线	有中线	无中线
线电压（V）	U_{AB}				
	U_{BC}				
	U_{CA}				
相电压（V）	U_{AN}				
	U_{BN}				
	U_{CN}				
电流（A）	I_A				
	I_B				
	I_C				

（2）无中线

断开中线，观察灯泡的亮度有无变化。重复测量各电压和电流值，记入表 3－2。

3．三相不对称负载星形连接

在图 3－2 中，A 相灯泡仍为 40W，B 相改为 2 只 40W 灯泡并联，C 相改为 2 只 60W 灯泡并联，仍按有中线和无中线两种情况测量各电压和电流值，填入表 3－2 中。

4．三相对称负载三角形连接

按图 3－3 接线，R_{AB}、R_{BC} 和 R_{CA} 都为 40W 灯泡，组成三角形接法，测量各电流值，填入表 3－3（电源电压为三相 220V 的线电压，以防烧坏灯泡）。

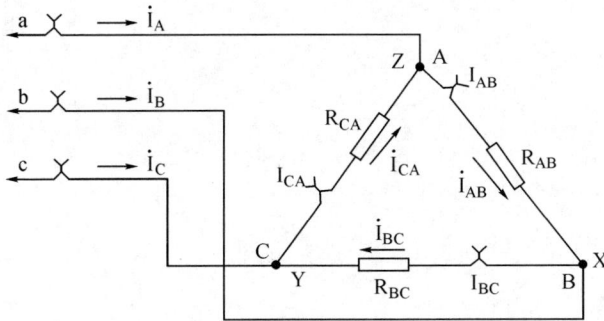

图 3－3

5．三相不对称负载三角形连接

将 R_{AB}、R_{BC}、R_{CA} 改为 40W、2×40W 和 2×60W 灯泡（注意相序不要弄错），再一次

测量各电流值，填入表 3-3 中。

表 3-3

电流负载	△ 负载对称			△ 负载不对称		
	A（40W）	B（40W）	C（40W）	A（40W）	B（2×40W）	C（2×60W）
相电流 I_P						
线电流 I_L						

【实验预习要求】

1. 掌握三相四线制电源系统有关内容。

2. 学会三相交流电路负载星形连接方法。

3. 学会三相交流电路负载三角形连接方法。

【实验报告要求】

1. 按规定内容和格式撰写报告。

2. 有一组灯泡其额定电压为 220V，功率为 100W，若接于"三相 380V 电源"时应如何接入？若接于"三相 220V"电源时又应如何接入，才能保证其正常工作。

3. 总结三相四线制中中线的作用。

4. 不对称三角形连接的负载，能否正常工作？实验中是否能证实这一点？

【实验仪器】

1. 三相灯板 1 套（4 块）。

2. 白炽灯泡 220V、40W 3 只，220V、60W 2 只。

3. 交流电压表 1 只。

4. 交流电流表 1 只。

3　Three-phase Circuit

Objectives

1. Learning the ways of connection of the Y- and Δ-connected circuit.
2. Examine the relations between quantities of lines and phases.
3. Understanding the function of neutral line.

Principle and Method

1. Source of three-phase four-wire system

For a balanced three-phase source, with a neutral line is available, the relations between line voltage U_L and phase voltage U_P are:

$$U_L = \sqrt{3}\,U_P$$

U_L: line voltage, the value of line-to-line voltage;

U_P: phase voltage, the voltage value between a line and the neutral wire

Usually, the voltage value of a three-phase four-wire system is effective value or rms value. So when we called "a three-phase 380V source", it means the line voltage is 380V and the phase voltage is 220V. In this experiment, the "three-phase 220V source" has line voltage of 220V and phase voltage of 127V.

2. Loads working under three-phase Y-connected source

(A) Balanced loads in a source of three-phase Y-connected, the line voltage U_L is $\sqrt{3}$ time of the phase voltage U_P, and the line current I_L equals the phase current I_P.

$$U_L = \sqrt{3}\,U_P$$

$$I_L = I_P$$

So, in this circumstance of balanced three-phase Y-connected source circuit, the current in the neutral line is zero, $I_0 = 0$, thus the neutral line can be omitted.

(B) If the loads connected with the three-phase Y-connected source is unbalanced, the circuit must be three-phase four-wire, of which the neutral line cannot be omitted. The existence of the neutral line can keep the regular voltage on the every line.

3. Loads working under three-phase Δ-connected source

Whether Loads working under three-phase Δ-connected source are balanced or not, the phase voltage is always equals line voltage:

$$U_L = U_P$$

(A) If the loads working under three-phase $\triangle$-connected source are balanced, the line currents are also balanced, the relation between phase current and line current is:

$$I_L = \sqrt{3}\,I_P$$

(B) If the loads working under three-phase $\triangle$-connected source are unbalanced, the line current does not equal $\sqrt{3}$ time of line current:

$$I_L \neq \sqrt{3}\,I_P$$

So far as the voltages U_L of the three-phase is symmetry, the phase voltages on the unbalanced loads are also symmetry, and the loads on every phase can work regularly.

Experiment

1. Measure the line voltage and phase voltage on a source of three-phase four-wire system

Because the filament lamps working as loads in this experiment have rated voltage of 220V, the three-phase 380V source has been changed into the three-phase 220V source by a three-phase voltage regulator.

Figure 3 – 1 shows the connection of a three-phase regulator.

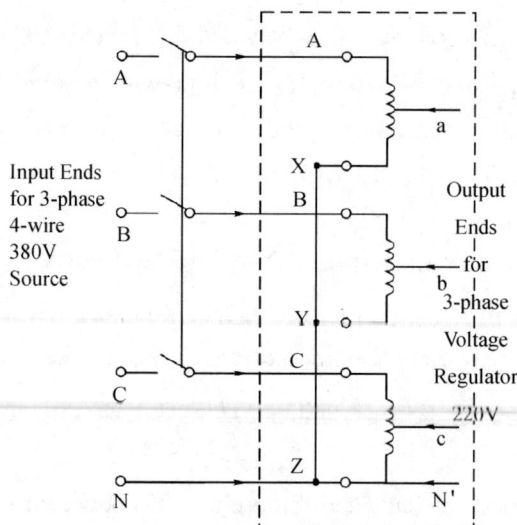

Figure 3 – 1

Measure the line voltage and phase voltage on a source of three-phase four-wire system, which is a "three-phase 220V source", and record the values in Table 3 – 1.

Table 3 – 1

Voltage	U_{AB}	U_{BC}	U_{CA}	U_{AN}	U_{BN}	U_{CN}
Three-phase 220V source						

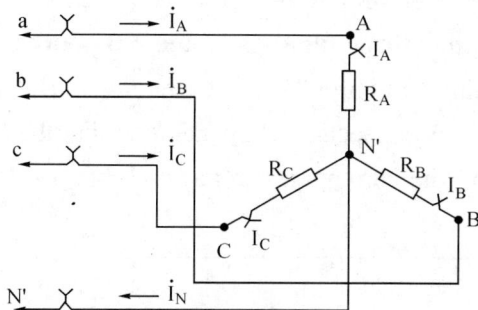

Figure 3 - 2

2. Balanced loads under three-phase Y-connected source

Three 40W filament lamps of are connected symmetrically under the source of three-phase Y-connected, as shown in Figure 3 - 2. Our measurement will go under the two cases: with a neutral line (N-line) and without neutral line. Keep the source in "three-phase 220V".

Table 3 - 2   Loads under Y-connection

		Balanced loads		Unbalanced loads	
		With N-line	Without N-line	With N-line	Without N-line
Line voltage (V)	U_{AB}				
	U_{BC}				
	U_{CA}				
Phase voltage (V)	U_{AN}				
	U_{BN}				
	U_{CN}				
Current (A)	I_A				
	I_B				
	I_C				

(A) With a neutral line.

Join the circuit. After checking by your instructor, switch on the three-phase source, and measure voltages and currents according to Table 3 - 2.

(B) Without neutral line

Break off the N-line; observe the brightness of the lamps. Measure voltages and currents again, and record them in Table 3 - 2.

3. Unbalanced loads under three-phase Y-connected source

In the circuit in Figure 3 - 2, keep the lamp 40W on phase A, change the lamp into two lamps of 40W in parallel on phase B, and change the lamp into two lamps of 60W in parallel on phase C. Measure corresponding voltage and current under the two cases of with a N-line and without N-line. Put your values in Table 3 – 2.

4. Balanced loads under three-phase Δ-connected source

Join the circuit according to Figure 3 – 3, where R_{AB}, R_{BC} and R_{CA} are all 40W filament lamps and connected in Δ-connection. Pay attention that the line voltage is 220V of the
three-phase source. Measure the currents according to Table 3 – 3.

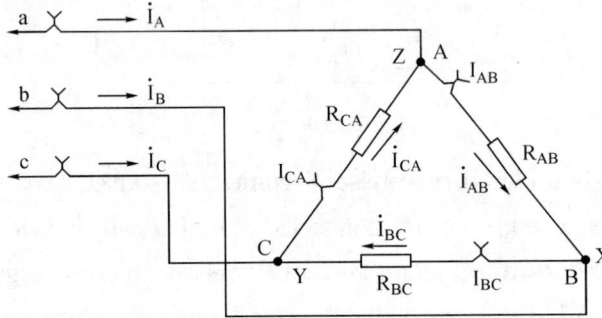

Figure 3 – 3

5. Unbalanced loads under three-phase Δ-connected source

Now change R_{AB}, R_{BC} and R_{CA} into 40W, 2×40W and 2×60W filaments, keep the
filament lamps in right phase order. Measure the current again and record them in Table
3 – 3.

Table 3 – 3

Loads	Balanced under Δ-connected			Unbalanced under Δ-connected		
	A (40W)	B (40W)	C (40W)	A (40W)	B (2×40W)	C (2×60W)
Phase current I_P (A)						
Line current I_L (A)						

Preparation Requirement

1. Mastering the knowledge about three-phase four-wire system.

2. Learning the way of connecting three-phase source in Y-connection.

3. Learning the way of connecting three-phase source in Δ-connection.

Report requirement

1. Write your report in format which is required in the preface.

2. There are some filament lamps which have 220V rated voltage and 100W rated power. For their regular working, how connect them into a source of "three-phase 380V", or
"three-phase 220V"?

3. Summarize working function of the N-line in the three-phase four-wire system.

4. Do the unbalanced loads in a Δ-connected source work regularly? How can you veri-

fy it?

Equipments

1. Special board of three-phase circuit, one set.
2. 220V 40W and 220V 60W filament lamps.
3. AC voltmeter.
4. AC ammeter.

实验四　三相异步电动机

【实验目的】

1. 了解异步电动机的铭牌数据。
2. 学习测量电动机绝缘电阻的方法。
3. 正确连接电动机的三相绕组，并使电动机启动和反转。

【实验原理与方法】

1. 电机的绝缘电阻

在使用电器设备时，其绝缘的好坏，对设备正常运行有密切的关系。绝缘材料的好坏用绝缘电阻的高低来衡量。由于设备受热、受潮等原因，会使绝缘电阻降低，甚至可能造成设备外壳带电和短路事故，所以在使用期间应做定期检查。

绝缘电阻大小可用兆欧表进行测量，一般是对绕组的相间绝缘及绕组与机壳之间的绝缘电阻进行测量，对于 500V 以下的中小型电机，其绝缘电阻最低不得小于 $1000\Omega/V$。

2. 对于鼠笼式三相异步电动机，当电源容量相对于电动机的功率是足够大时，一般采用直接启动法，即将电动机的定子绕组直接接入电网，加以额定电压直接启动。

异步电动机的旋转方向，取决于定子旋转磁场的旋转方向，而旋转磁场的旋转方向又取决于三相电源接入的相序。故只要改变三相电源与定子绕组连接的相序，即可使电动机改变旋转方向。

3. 图 4-1 为电动机出线盒上的接线板，三相绕组的六个线头分别接在六个接线柱上，可进行三相绕组极性判别实验与直接启动实验。当绕组极性确定后，三个绕组六个绕头按图中标号连接。

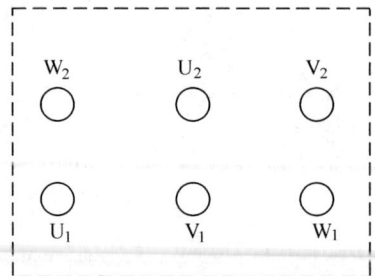

图 4-1

根据需要可将定子绕组连接成 Y 形或△形两种方式。目前我国生产的三相异步电动机额定电压一般为 380V。定子绕组的连接方式根据电动机容量大小分成两类，额定功率在 4kW 以下的小容量电动机采用星形连接；额定功率在 4kW 以上的电动机采用三角形连接。两种连接方式见图 4-2。

【实验内容与步骤】

1. 用手拨动转子，细听电动机内部有无摩擦声，检查轴承润滑情况，传动装置是否良好。

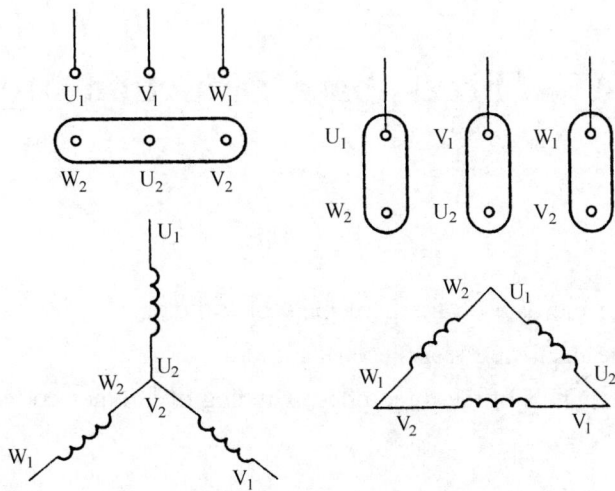

图 4 - 2 三相绕组星形与三角形连接方式

2．记录铭牌数据（电动机铭牌识别见实验五）。

型 号 ＿＿＿＿＿＿＿功 率 ＿＿＿＿＿＿＿接 法 ＿＿＿＿＿＿＿电 压 ＿＿＿＿＿＿＿电 流

＿＿＿＿＿＿＿转速＿＿＿＿＿＿＿

3．用兆欧表测量绝缘电阻（附录三《兆欧表的使用方法》）。

4 AC Three-phase Induction Motor

Objectives

1. Understanding the data on the nameplate of a motor.

2. Measuring the insulating resistance of a motor.

3. Joining the terminals of the three-phase winding of a motor correctly to start or to reverse a motor.

Principle and Method

1. Insulating resistance of an electric motor.

The insulating capability about electric equipment has an important relation with the working performance of it. The insulating capability of the insulating material is evaluated by its insulating resistance value. Working in a heat or damp surrounding, the insulating resistance of the equipment can be reduced, and indeed it can cause the shell of the equipment electrification or short circuit accident. So the insulating resistance should be checked regularly.

The value of insulating resistance can be measure by a meg-ohm meter. Ordinarily, we will check the resistance between the phases of the motor, and the resistance between windings and shell. For the medium and small power motor of 500V (rating voltage) or lower, the insulating resistance should not be less than 1000 Ω/V.

2. For AC three-phase cage motor, if the source capability is big enough comparing with the rating power, usually we start up the motor directly, that is we join the stator windings with the electric source to get the rating voltage directly.

The rotating direction of an induction motor is determined by the rotating magnetic field produced by the stator. And the rotating direction of rotating magnetic field is determined by the joining phase sequence of source. Therefore we can rotate the motor in reverse by changing the joining phase sequence of the source.

3. Figure 4 − 1 is a schematic drawing of the wire connection board leading from windings of a motor. The six wire heads from three windings connect with the six

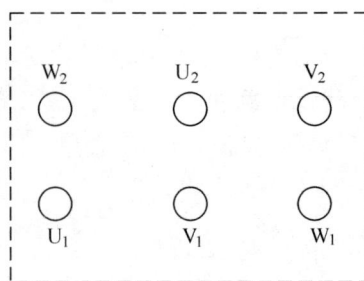

Figure 4 − 1

posts, by which we can judge the poles and start up the motor. After the polarities have been determined, the six wire heads can be joined according to the marks in on the board.

There are two ways of joining the stator windings: Y-connection and Δ-connection, which should be choose according to special requirement. The rating voltage of the AC three-phase induction motor in our country is 380V usually. The two joining ways of the stator windings can be determined by the power capability: motors of rating power smaller than 4kW can be Y-connected and the others can be Δ-connected. The two joining ways are shown in Figure 4 – 2.

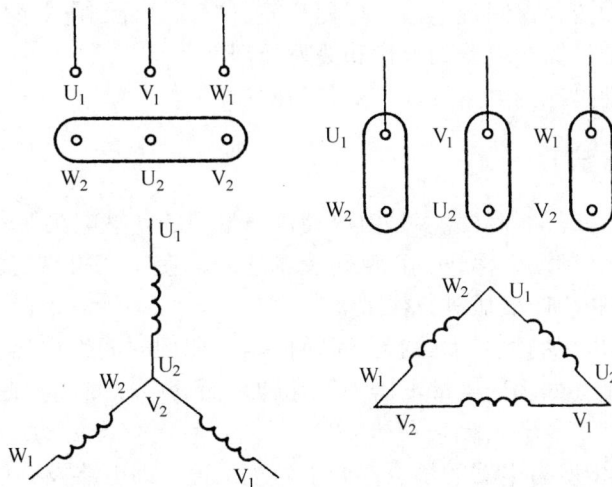

Figure 4 – 2   Y-connection and Δ-connection

Experiment

1. Check the motor: dial the rotor to hear if there is abnormal sound; feel the lubrication of the bearings; exam the gear-drive equipment.

2. Record the data on the nameplate of the motor. (The signification of the nameplate can be seen in next Experiment.)

Type _____   Rating Power _____ kW

Connecting way _____   Rating Voltage _____ V

Rating current _____ A   Rotating speed _____ r/min

3. Measure the insulating resistance by a meg-ohm meter. (See Appendix 3: Operation Guide of Meg-ohm Meter)

实验五　异步电动机的继电接触器控制线路（设计性实验）

【实验目的】

1. 了解交流接触器、时间继电器、行程开关、按钮等控制电器的结构及其使用方法。
2. 学习异步电动机控制电路的连接和故障处理。
3. 学会异步电动机控制电路设计、安装和调控。

【实验原理与方法】

企业生产过程中，运动部件大多是由电动机来带动的。继电接触器控制电路目前大量应用于电动机的启动、制动、停止、正反转及调速等，使生产机械能按既定的顺序动作，同时，也能对电动机和生产机械进行保护。

对电动机和生产机械的控制大致有：时间控制（如用时间继电器）；行程控制（如用行程开关）；电流控制（如用速度继电器）等几种。而其中以时间控制和行程控制应用得比较广泛。

控制线路原理图中所有电器的触点都处于静止位置，即电器没有任何动作的位置。例如，对于接触器或继电器来说，就是线圈没有电流时的位置；又如按钮就是没有受到压力的位置。

【实验内容和步骤】

1. 异步电动机单方向启动—停止的控制

（1）了解交流接触器、热继电器、按钮等控制电器的结构及其动作原理。

（2）用万用表的电阻档查明在实验中所用上述电器的常开触点、常闭触点和线圈所对应的接线柱。

在断开电源的条件下，用手将控制按钮反复按下、松开，并用万用表欧姆档检查常开、常闭触点的接触情况（应无接触电阻）。再检查线圈的电阻值是否正常。

（3）按图 5 – 1 接线。

三相异步电动机单向启动与停止控制电路，要求按下启动按钮（发出启动指令）后，电动机能长时间连续运行。当按下停止按钮时，电动机停止运转。

该电路的工作原理是：当闭合刀开关 Q 并按下启动按钮 SB_1 后，控制电路经常闭按钮 SB_2 接通接触器励磁线圈 KM，使主触点 KM 吸合，接通主电路，使电动机启动运行。与此同时，接触器的常开辅助触点 KM 亦被吸合。这时就可以松开启动按钮 SB_1，控制电路通过辅助触点 KM 保持通电，维持接触器主触点闭合电动机继续运行。这种利用接触器本身的常开辅助触点闭合，以维持接触器线圈持续通电的作用称为自锁，又称自保护。起自锁

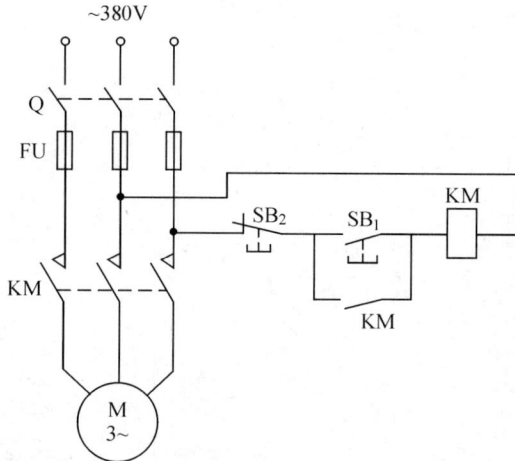

图 5 - 1　启动和停止控制电路原理图

作用的触点称为自锁触点。要使电动机停止运行，必须在控制电路中加入停止按钮 SB_2（常闭）。当电动机依靠自锁作用连续运行时，只需按下 SB_2，即可断开控制电路，使接触器线圈断电，其主触点断开。与此同时，常开辅助触点也断开。此后，即使松开停止按钮 SB_2，控制电路也不可能再接通，电动机保持停止状态。

2．异步电动机的时间延时控制

按图 5 - 2 连接。本实验中，交流接触器 KM_1 和 KM_2 分别控制两台电动机，第 2 台电动机在第 1 台启动后，通过时间继电器 KT 使其延时 Δt 秒启动。实验电路中，省去了第 1 台电动机，用 KM_1 的吸合声代替第 1 台电动机的启动。

该控制电路的工作原理和动作顺序是

按下启动按钮 SB_1 → 接触器 KM_1 线圈通电

→ { 主触点 KM_1 闭合 → M_1 电动机启动

常开辅助触点 KM_1 闭合 → 自锁

时间继电器 KT 线圈通电→

→延时→常开触点 KT 延时闭合 → 接触器 KM_2 线圈通电→

→ { 主触点 KM_2 闭合 → M_2 电动机启动

常开辅助触点 KM_2 闭合 → 自锁

常闭辅助触点 KM_2 断开 → KT 线圈通电→常开触点 KT 断开

（避免 KT 长时间通电，消耗电功率）

3．异步电动机的行程模拟控制

用行程开关控制电动机正转和反转，模拟控制工作台的前进与后退。这个电路要实现的功能如下：电动机一经启动，机械部件就无需人的操作而自动在 A、B 两个位置之间往返运动，直到按下停止按钮 SB_2 时才停止。

电路如图 5 - 3 所示，该控制电路实际上就是一个三相异步电动机的正反转控制电路。当电动机带动生产机械正向运行至某一极限位置 A 时，应能够自行停车。为此在 A 处放置一个行程开关 ST_A，且将它的一对常闭触点串联接入正转接触器 KM_F 线圈控制电路内。

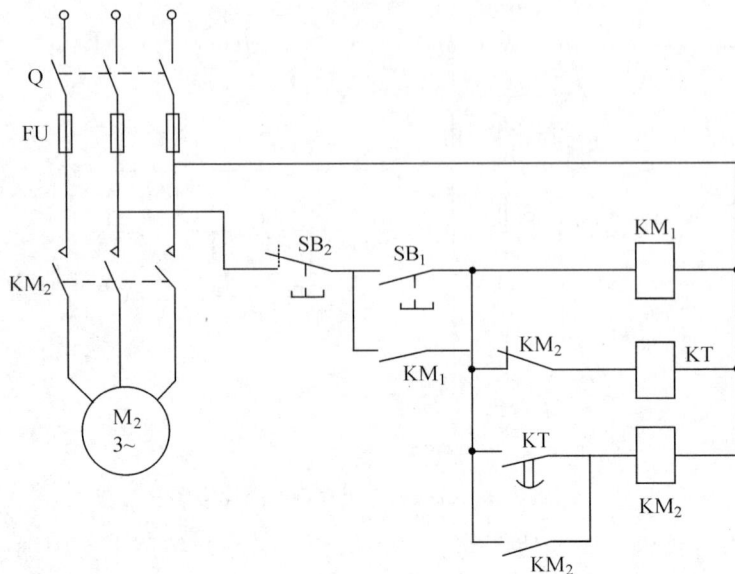

图 5 – 2   采用时间继电器的延时启动控制电路

同样在机械部件反向运行的某一极限位置 B 处放置行程开关 ST_B，且将它的一对常闭触点串连接入反转接触器 KM_R 线圈控制电路中。欲使该电动机正转、机械部件正向运行，则按下启动按钮 SB_F。当机械部件运行到极限位置 A 时，装在机械部件上的挡块触压行程开关 ST_A，使其常闭触点断开，并将 KM_F 线圈断电、电机停转。电动机反转、机构部件反向运行时的限位控制与上述情况相似。如我们再将行程开关中反向行程开关 ST_B 的常开触点与正转启动按钮 SB_F 并联，将正向行程开关 ST_A 的常开触点与反转启动按钮 SB_R 并联，则可使电动机交替正、反转运行，机构部件在 A、B 位置之间自动往复运动。

　　该电路的动作顺序是：若机械部件处于 A、B 位置之间的任一点处，则：

按下启动按钮 SB_F → 接触器 KM_F 线圈通电→

→$\left\{\begin{array}{l}\text{主触点 } KM_F \text{ 闭合} → \text{电动机正转} \\ \text{常开辅助触点 } KM_F \text{ 闭合} → \text{自锁} \\ \text{常闭辅助触点 } KM_F \text{ 断开} → \text{互锁}\end{array}\right\}$→电动机带动机械部件正向运行……

→到位置 A 触压行程开关 ST_A→

→$\left\{\begin{array}{l}\text{常闭触点 } ST_A \text{ 断开} → KM_F \text{ 线圈断电} → \left\{\begin{array}{l}\text{电动机 M 停止正转} \\ \text{常闭辅助触点 } KM_F \text{ 闭合为 } KM_R \text{ 线圈通电创造条件}\end{array}\right. \\ \text{常开触点 } ST_A \text{ 闭合} → KM_R \text{ 线圈通电}→\end{array}\right.$

→$\left\{\begin{array}{l}\text{主触点 } KM_R \text{ 闭合} → \text{电动机反转} \\ \text{常开辅助触点闭合} → \text{自锁} \\ \text{常闭辅助点断开} → \text{互锁}\end{array}\right\}$→电动机带动机械部件反向运动……

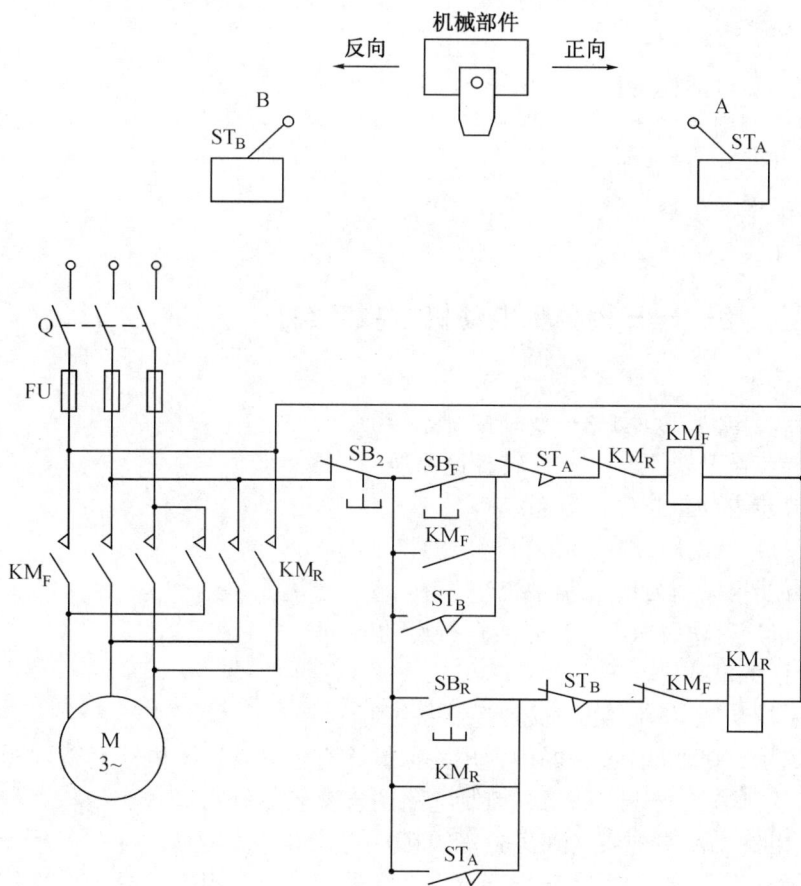

图 5 – 3 行程控制电路

【实验预习要求】

1．弄清控制电路中各图形符号和文字符号的意义及其在电路中的作用。

2．读图 5 – 1，说明按钮 SB_1 两端并联的常开辅助触点 KM 的作用。

3．交流接触器的线圈额定电压为 380V。若误接到交流 220V 电源上，会产生什么后果？为什么？

【实验报告要求】

1．回答图 5 – 3 异步电动机的行程控制线路中，辅助常闭触点 KM_R 和 KM_F 的作用是什么？

2．阐述用万用电表检查图 5 – 2 控制电路的方法。

3．回答预习要求中"2"、"3"提出的问题。

4．试列表说明图 5 – 3 控制电路中各控制电器的文字符号、图形符号、静态位置及工作状态的位置。

【实验仪器】

1．三相异步电动机 1 台。

2．交流接触器 2 只。

3．行程开关 2 只。

4．时间继电器 1 只。

5．按钮 4 只。

【设计型实验——三相异步电动机控制系统】

1．设计选题：

①三相异步电动机点动——连续控制系统

②三相异步电动机正转——反转控制系统

③三相异步电动机顺序控制系统

      （M_1 启动后，M_2 才能自行启动）

④三相异步电动机延时控制系统

      （M_1 启动后 10s，M_2 才能自行启动）

2．设计任务书：

①控制系统功能（例：正 – 反转控制系统）。

②设计要求：（例：有过载保护、短路保护、有电气互锁、机械互锁）。

3．画出控制系统——主电路、控制电路图（交老师审阅）。

4．列出电器元件明细表（到实验室领料）。

（例：名称　　　　　文字符号　　　　型号　　　　规格　　　　数量）

    交流接触器　　　KM_1　　　CJ10 – 10　　　380V/10A　　　2

5．连接调试控制系统。

6．指导老师评定打分。

5　Relay-contactor Control Circuit for AC Three-phase Induction Motor

Objectives

1. Understanding the structure and operation of control circuit for AC contactor, time-delay relay, travel switch and button.

2. Learn to join the control circuit for AC induction motor and deal with the circuit malfunction.

3. Learn to design, install and adjust circuit for AC induction motor.

Principle and Method

In a manufactory most of the moving parts are driven by electric motors. Relay-contactor control circuit are been applied to starting, braking, stopping, normal-reverse transferring, speed governing and so on. Relay-contactor control circuit can not only drive machines working according to certain sequence, but also protect motors and machines properly.

There are certain kinds of controlling, such as time control (by time-relay), position control (by position switch or travel switch), and electricity control (by speed relay). And time control or position control are most commonly used.

Note that all the contact points in the circuit schematic diagram are at rest positions, of which there is no action. For example, the rest positions are the positions with the coils under no electricity for a contact or relay, or button under no press.

Experiment

1. Control the one direction start-stop for an induction motor.

(A) Knowing the structure and action principle of AC contactor, thermal relay and button.

(B) Check the normally open contact, normally close contact and the wiring posts by ohm step of your multimeter.

Under the condition of power off, press and release the button to check the resistance of the button switch. The normally open contact and normally close contact should have correct resistance values.

(C) Join the circuit according to Figure 5 – 1.

For the circuit of one direction start-stop for an AC induction motor, if the start button is pressed (it a start order.) the motor can works normally and continuously, if the stop button is pressed the motor stops.

The working sequence of this circuit is: after close the knife switch Q and press the start button SB_1, the excitation coil KM is power on through the normally close button SB_2 of the control circuit, and the main circuit is turn-on and the motor start. At the same time the normally open auxiliary contact of KM is closed. Now the button SB_1 can be released, the control circuit is keeping on power and the motor can work continuously. This process to keep the contactor on power by the closing the normally open auxiliary contact is self-hold or self-lock. To stop the motor, there should be a stop button SB_2, which is normally closed. When the motor works continuously with the self-lock function, press SB_2 to break the control circuit, and coil of the contactor power off to open the main contacts, meanwhile the normally open auxiliary contacts open. After that, release the button SB_2, the control circuit open and the motor keeps out of working.

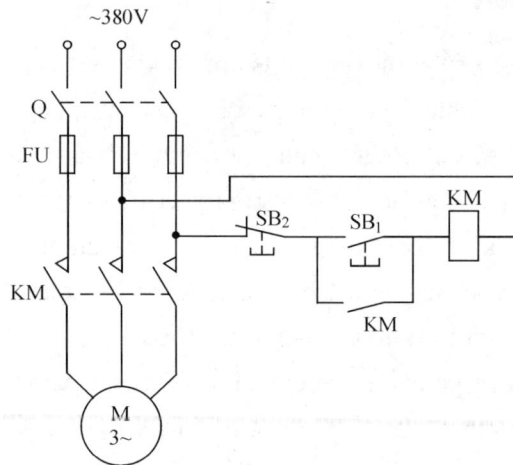

Figure 5 – 1   Start-stop control circuit

2. Timing control of AC induction motor

Join the circuit in Figure 5 – 2. In this experiment, AC contactor KM_1 and KM_2 control two motors separately, the second motor should be started in a time of Δt after the first motor start by time relay KT. In this circuit, the first motor is omitted, and its starting is indicated by the attraction sound of KM_1.

The working sequence of this circuit is:

Press the start button SB_1 → Power on the coil of contactor KM_1

→ {
Close the main contact of KM_1 → Start Motor M_1

Close the normal open auxiliary contact of KM_1 → Self-lock

→ Power on time relay KT → Delay → Close the normal open contact of KT
}

$\rightarrow$ Power on the coil of KM$_2$ $\rightarrow$

$\left\{\begin{array}{l}\text{Close the main contact of KM}_2 \rightarrow \text{Start Motor M}_2 \\ \text{Close the normal open auxiliary contact of KM}_2 \rightarrow \text{Self-lock} \\ \text{Open the normal close auxiliary contact of KM}_2 \rightarrow \text{Power off KT} \rightarrow \text{Open KT}\end{array}\right.$

(It is unnecessary to keep KT on power for a long time to consume electric power.)

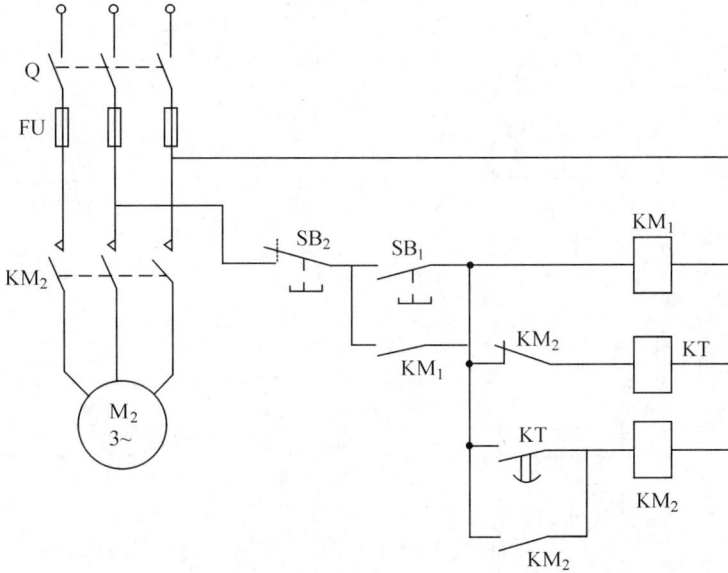

Figure 5 – 2   A time delay circuit controlled by a time relay

3. Simulation the travel control of AC induction motor.

The process of controlling motor positive or and reverse rotation by travel switch can simulate the controlling of the forward and back step of a working platform. The function of the circuit can be described as following: once the motor has been started, certain machine parts can move between position A and B without any additional operation until the stop button SB$_2$ is pressed down.

The circuit is shown in Figure 5 – 3. Briefly speaking, this circuit is a controlling circuit for an AC induction motor rotating positively or reversely. When a machine part driving by the motor moves to a certain limit position A, it can stop automatically. So, a travel switch ST$_A$ has been set at position A, and its normal close contact pair has been joined in series in the circuit of coil KM$_F$, which can control the motor rotate positively. Similarly, set a travel switch ST$_B$ at position B, which is the limit position of back moving of the machine part, and its normal close contact pair has been joined in series in the circuit of coil KM$_R$, which can control the motor rotate reversely. Now, press the SB$_F$, the motor rotates forward and the machine part moves forward. And when the machine part moves to limit position A, the baffle plate will press the travel switch ST$_A$ to open the normal close contact, and cut the power of KM_F coil, the motor stops. In the same way, the motor and the machine part can be controlled to rotate backward. If the normal open contact of ST_B is joined in parallel with the forward staring button SB_F and the normal open contact of ST_A is joined in parallel with the backward staring button SB_R, the back and forward rotating of the motor will be alternate, and the machine part will move reciprocally between position A and B.

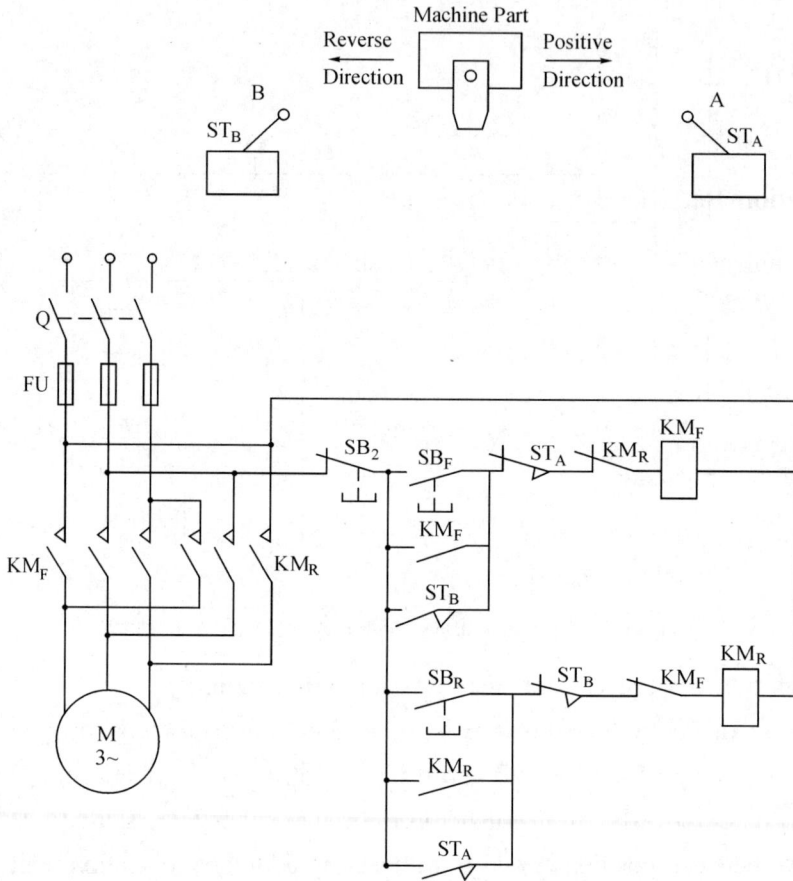

Figure 5 – 3 Traveling control circuit

The working sequence of this circuit is: if the machine part which is going to be driven by the motor is placed between position A and B, the working process will be the following:

Press start button $SB_F \rightarrow$ Power on the coil of contactor $KM_F \rightarrow$

$\rightarrow \begin{cases} \text{Close the main contact of } KM_F \rightarrow \text{Motor rotate forward} \\ \text{Close auxiliary normal open contact of } KM_F \rightarrow \text{Self-lock} \\ \text{Open auxiliary normal close contact of } KM_F \rightarrow \text{Interlock} \end{cases} \rightarrow \text{Motor drives machine}$

part $\cdots \rightarrow$ Press travel switch ST_A at position A $\rightarrow$

Open normal close contact of ST_A → Power off KM_F ┌→ Motor stops

└→Close auxiliary normal close contact of KM_F, which is the condition for electrify the coil of KM_R

Close normal open contact of ST_A → Power on KM_R→

Close the main contact of KM_R→ Motor rotate reversely

Close auxiliary normal open contact of KM_R→Self-lock } → Motor drives machine part reversely···

Open auxiliary normal close contact of KM_R→Interlock

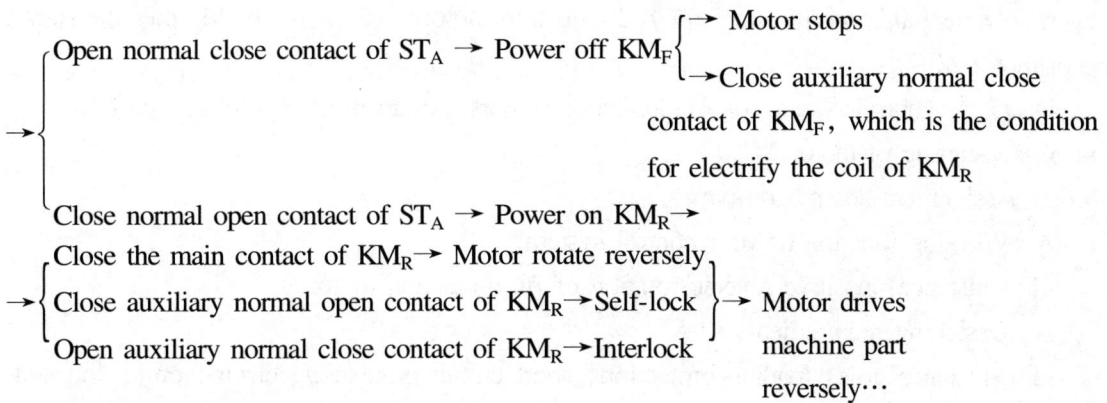

Preparation Requirement

1. Be familiar with the drawing signs and letter symbols in the circuit, understand their signification and function.

2. According to Figure 5 – 1, explain the function of the auxiliary normal open contact of KM, which is in parallel with button SB_1.

3. The rating voltage of AC contactor is 380V. If it is joined with a source of 220V, what will happen in this experiment? Why?

Report Requirement

1. According to the traveling control circuit in Figure 5 – 3, explain the function of the auxiliary normal close contacts of KM_R and KM_F.

2. Illustrate the detail of the method of checking the circuit in Figure 5 – 2.

3. Answer the questions 2 and 3 in the "Preparation Requirement".

4. List all the drawing signs and letter symbols, and give their normal and working positions.

Equipments

1. Three-phase AC induction motor

2. AC contactor, 2 sets

3. Travel switch, 2 sets

4. Time-delay relay

5. Button, 4 sets

Designing Experiment – AC Induction Motor Control System

1. Select a designing subject:

A. Control system for discontinuous-continuous rotation of AC induction motor.

B. Control system for positive-reverse rotation of AC induction motor.

C. Sequential control system for AC induction motors. (Start motor M_2 after the starting of motor M_1.)

D. Delay control system for AC induction motors. (Start motor M_2 in a time of 10 second of the starting of motor M_1.)

2. Assignment application paper:

A. Working function of your control system.

(For instance: positive-reverse rotation of AC induction motor.)

B. Designing requirement.

(For instance: over loading protection, short circuit protection, electric protection and mechanism protection.)

3. Control circuit figure, including main circuit and controlling circuit, which should be checked and approved by your instructor.

4. List your requiring electric elements, which will get from lab.

For instance:

Title	Letter Symbol	Type	Spec	Amount
AC Contactor	KM_1	CJ10 – 10	380V/10A	2

5. Join your control system circuit and adjust.

6. Your lab instructor will assess your job and give a score.

实验六 常用电子仪器的使用

【实验目的】

1. 了解双迹示波器的工作原理，熟悉使用方法。
2. 学会正确使用常用电子仪器。

【实验原理与方法】

1. 常用仪器仪表

在模拟电子电路基础实验中，常用的电子仪器有如图 6-1 所示的相关仪器，其相互关系及作用如下：

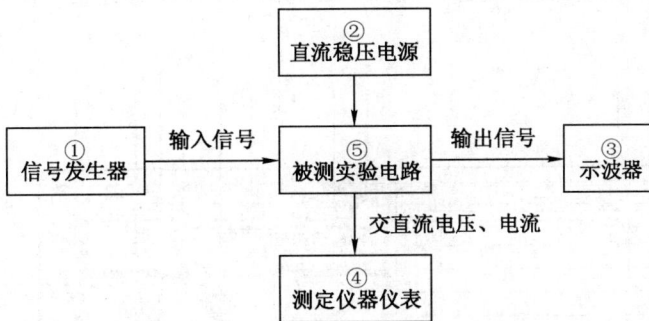

图 6-1 模拟电路实验常用仪器仪表的相互关系

（1）信号发生器：用来产生信号源的仪器，它有正弦波、三角波、方波输出，输出电压和频率均可调节。有数字表或指针表指示其输出电压大小，输出波形根据被测实验电路要求进行选择。

（2）直流稳压电源：为被测实验电路提供能源，通常是电压输出，例如 5~6V，±12V或±15V，交流双~15V或单~9V等。

（3）示波器：用来观察实验电路的输出信号。通过示波器可显示电压或电流波形，可测量频率、周期及其他有关参数。

（4）测量仪器仪表：这主要是指测量实验电路中的电阻、电压、电流、频率等参数的常用仪表，例如晶体管毫伏表、电流表、指针式万用表、数字万用表、频率计等。

（5）被测实验电路：它是研究模拟电路的基础，通过相关仪器准确地测量数据，观察实验现象和结果，进而真正掌握该电路的作用。

2. 双迹示波器

双迹示波器可同时观察和测定两种不同信号的瞬变过程，它不仅可以把两种不同的电信号同时在屏幕上显示，以提供对比、分析，测定其相互关系与相关参数，还可使两组信号构成差分放大的形式，或使两信号迭加后显示。此外，这种示波器也可只用一个通道工作，作为通常的单迹示波器。

双迹示波器和普通单迹示波器相比，最主要的特点能双迹显示。产生双迹显示有两种方法：一种是利用双束射线示波管；另一种是利用电子开关。CS－4125 示波器是利用电子开关产生双迹显示。

图 6－2 为电子开关简图。其工作原理如下：借助于电子开关 S，通道 Y_A 和 Y_B 的外来信号交替地耦合至 Y 轴后级放大器，电子开关的转换频率应足以使人眼将两幅仅在特定瞬间出现的图形看成是同时的波形。

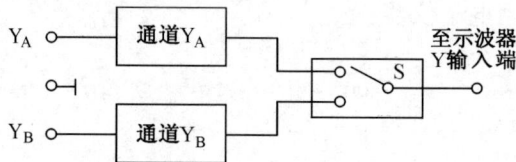

图 6－2

图 6－3 是 CS－4125 双迹示波器的前面板图，各控制件的功能和作用分述如下：

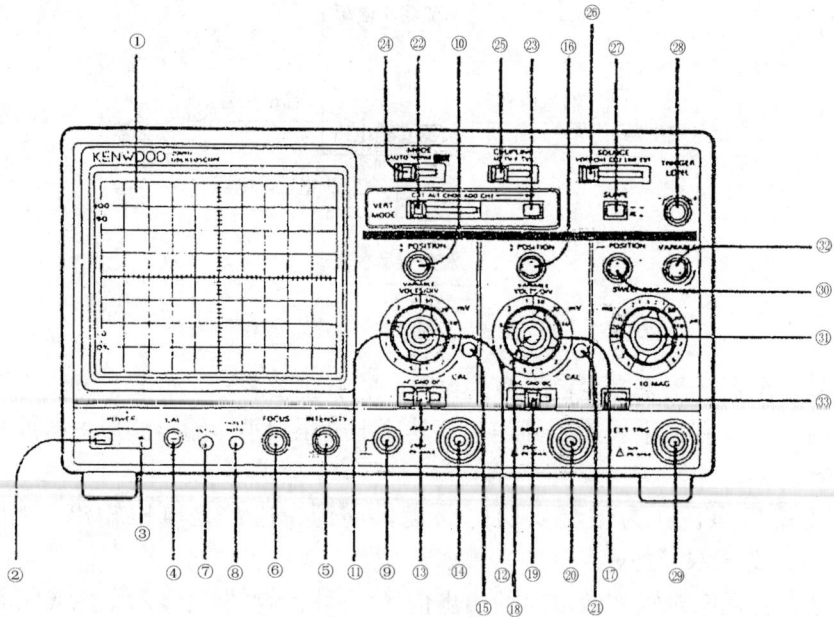

图 6－3

序号	控制件名称	功能
(1)	亮度（INTENSITY）	调节光迹的亮度
(2)	辅助聚焦（ASTIG）	与聚焦配合，调节光迹的清晰度
(3)	聚焦（FOCUS）	调节光迹的清晰度
(4)	迹线旋转（TRACEROTA）	调节光迹与水平刻度线平行
(5)	校正信号（CAL）	提供幅度为 $1V_{P-P}$，频率为 1kHz 的方波信号，用于校正 10:1 探极的补偿电容器和检测示波器垂直与水平的偏转因数

序号	控制件名称	功能
(6)	电源指示（POWER INDICATOR）	电源接通时，灯亮
(7)	电源开关（POWER ON/OFF）	电源接通或关闭
(8)	CH 1 移位（POSITION）	调节通道 1 光迹在屏幕上的垂直位置
(9)	CH 2 移位（POSITION）	调节通道 2 光迹在屏幕上的垂直位置
(10)	垂直方式（VERTICAL MODE）	CH 1 或 CH 2：通道 1 或 2 通道单独显示 ALT：两个通道交替显示 CHOP：两个通道断续显示，用于扫速较慢时的双踪显示 ADD：用于两个通道的代数和或差
(11)	CH 1 电压衰减器（VOLTS/DIV）	CH 1 通道垂直偏转灵敏度
(12)	CH 2 电压衰减器（VOLTS/DIV）	CH 2 通道垂直偏转灵敏度
(13)	CH 1 微调（VARIABLE）	用于连续调节垂直偏转灵敏度，顺时针旋到底为校正位置，可定量测量信号幅度
(14)	CH 2 微调（VARIABLE）	用于连续调节垂直偏转灵敏度，顺时针旋到底为校正位置，可定量测量信号幅度
(15)	CH 1 耦合方式（AC – GND – DC）	用于选择 CH 1 通道被测信号馈入垂直通道的耦合方式
(16)	CH 2 耦合方式（AC – GND – GC）	用于选择 CH 2 通道被测信号馈入垂直通道的耦合的方式
(17)	CH 1 INPUT	被测信号的输入插座；通道 1
(18)	CH 2 INPUT	被测信号的输入插座；通道 2
(19)	接地（GND）	与机壳相联的接地端
(20)	外触发输入（EXT INPUT）	外触发输入插座
(21)	内触发源（INT TRIG SOURCE）	用于选择 CH 1，CH 2 或交替触发
(22)	触发源选择（TRIG SOURCE）	用于选择触发源 INT（内），EXT（外）或 LINE（电源）
(23)	触发极性（SLOPE）	用于选择信号的上升或下降沿触发扫描
(24)	电平（LEVEL）	用于调节被测信号在某一电平触发扫描
(25)	微调（SWEEP VARIABLE TIME）	用于连续调节扫描速度，顺时针旋足为校正位置，可定量测量信号周期
(26)	扫描速率（SWEEP TIME/DIV）	用于调节扫描速度
(27)	触发方式（TRIG MODE）	常态（NORM）：无信号时，屏幕上无显示，有信号时，与电平控制配合显示稳定波形 自动（AUTO）：无信号时，屏幕上显示光迹；有信号时，与电平控制配合显示稳定波形 电视场（TV）：用于显示电视场信号 峰值自动（P – P AUTO）：无信号时，屏幕上显示光迹；有信号，无须调节电平即能获得稳定波形显示
(28)	触发指示（TRIG'D）	在触发扫描时，指示灯亮
(29)	水平移位（POSITION PULL × 10）	调节迹线在屏幕上的水平位置 拉出时扫描速度被扩展 10 倍

3．波形的观察

（1）将有关控制件按下表设置

控制件名称	作用位置	控制件名称	作用位置
亮度（INTEN）	居中	触发方式	AUTO
聚焦（FOCUS）	居中	扫描速度（TIME／DIV）	0.5ms
位移（CH1，CH2）	居中	极性（SLOPE）	正
垂直方式（MODE）	CHOP	触发源	INT
电压衰减器（VOLTS／DIV）	1V	内触发源	CH1
微调（VARIABLE）	校正位置	输入耦合	AC

（2）接通电源，电源指示灯亮，稍候预热，屏幕上出现光迹，分别调节亮度、聚焦、辅助聚焦、垂直移位、水平移位、迹线旋转，使光迹清晰居中并与水平刻度平行。

（3）垂直方式的选择

当只需观察一路信号时，将"MODE"开关置"CH1"或"CH2"，此时选中的通道有效，被测信号可从通道端口输入。当需要同时观察两路信号时，将"MODE"开关置交替"ALT"，该方式使两个通道的信号被交替显示，交替显示的频率受扫描周期控制。当扫描低于一定频率时，交替方式显示会出现闪烁，此时应将开关置于数据断续"CHOP"位置。当需要观察两路信号代数和时，将"MODE"开关置于"ADD"位置，在选择这种方式时，两个通道衰减设置必须一致，CH2 移位处于常态时为 CH1 + CH2，CH2 移位拉出时（PULLINVERT）为 CH1 – CH2。

（4）扫描速度的设定

扫描范围从 0.2μs/DIV ~ 0.5S/DIV 按 2 倍、5 倍进位分 20 档，微调提供至少 2.5 倍的连续调节，根据被测信号频率的高低，选择合适的档级，在微调顺时针旋足至校正位置时，可根据开关的指示值和波形在水平轴方向上的距离读出被测信号的时间参数，当需要观察波形某一个细节时，可进行水平扩展 ×10，此时原波形在水平轴方向上被扩展 10 倍。

4．交流电压的测量

从 y 轴的一个通道输入正弦波，调节"扫描速度"、"扫描微调"使显示波形稳定，并将该通道的"微调"右旋到底。调节"Y轴灵敏度"，使波形幅度合适，读出幅度的 div 数 H，和此时的 V/div 值，按下列公式求出它的峰–峰值 U_{P-P}，幅值 U_m 和有效值 U（图 6–4 中 $H = 7$）。

波形幅度上峰点和谷点之间 7 格，根据 y 轴灵敏度开关"V/div"所指为 0.2V/div（即每格 0.2V）可推算出电压峰–峰值，按下列关系可换成有效值 U，即：

$$U_{P-P} = 0.2 \times 7 = 1.4V \qquad U = \frac{U_{P-P}}{2\sqrt{2}} = \frac{1.4}{2.8} \text{ V} = 0.5V$$

波形两个谷点（或两个峰点）间的水平距离 2 格，根据扫描速度选择开关"t/div"所指 0.5ms/div（即每格 0.5ms），可推算信号电压的周期为：

$$T = 0.5 \times 2\text{ms} = 1\text{ms}$$

其频率为：

$$f = \frac{1}{T} = \frac{1}{1 \times 10^{-3}} \text{ Hz} = 1000\text{Hz} = 1\text{kHz}$$

$$U_{P-P} = \text{V/div} \times H\text{div}$$

$$U_m = \frac{1}{2} U_{P-P}$$

$$U = \frac{U_m}{\sqrt{2}}$$

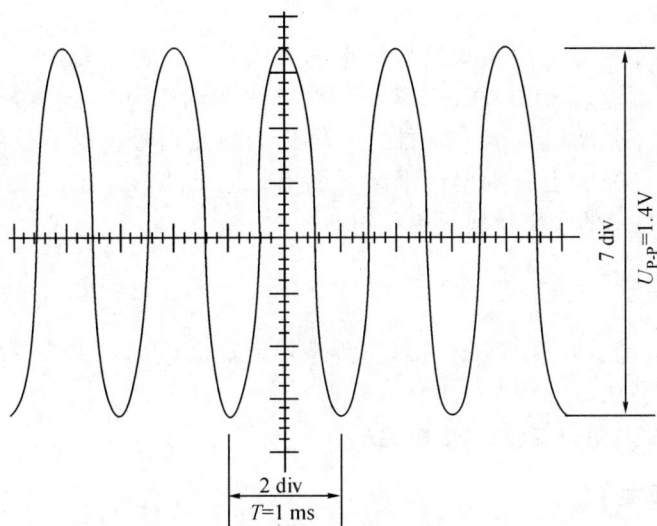

图 6 – 4　0.5V、1kHz 信号电压波形

5. 频率测量

（1）直接测量法

从 y 轴的一个通道先后输入正弦波、方波和锯齿波，将"扫描微调"置"校准"即右旋到底，调节"扫描速度"，使示波器先后显示 2 个正弦波，3 个方波，4 个锯齿波，分别读出它们的波形宽度的 div 数 L 和对应的 t/div 值，按下列公式求出它们的 T 和 f：

$$T = \text{t/div} \times L\text{div}$$

$$f = \frac{1}{T}$$

（2）李萨如图形法

李萨如图形法适于测量较低频率的信号，其方法是：在示波器的 x 轴和 y 轴分别输出两个简谐信号（即正弦波），其中一个是已知频率的信号，一个是待测信号（通常加于 y 轴），荧光屏上会显示出李萨如图形，但对应不同的相位和信号的频率比，图形不同，且当频率是简单的整数比时，荧光屏上能形成一个简单清晰的图形。图 6 – 6 是频率比 f_y/f_x 为 1、2、3 三种情况下，某一相位差下得到的李萨如图形。如果在荧光屏表面水平和垂直方向上作两条互相垂直的线，数出水平直线与李萨如图形的切点数 N_x 和垂直直线与李萨如图形的切点数 N_y，则这两个切点数之比就是信号的频率比，由此可求出未知信号的频率。即：

$$f_y = f_x \cdot \frac{N_x}{N_y}$$

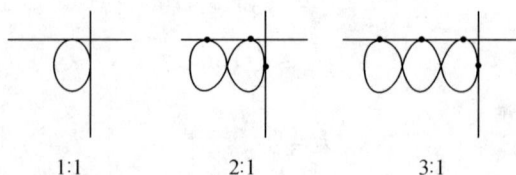

| 1:1 | 2:1 | 3:1 |

图 6 - 5

本实验用李萨如图形法测未知信号频率的方法如下：将〔MODE〕置于"X - Y"ON位置。从信号发生器输入 50Hz 的正弦波至示波器的 x 轴作为标准信号，从另一信号发生器输入未知频率的正弦波信号至示波器的 y 轴作为被测信号，调节被测信号的频率及示波器有关旋钮，使荧光屏上分别出现李萨如图形，记录波形并求出被测信号对应的频率 f_y，和信号发生器显示的频率比较是否相符。

【实验预习要求】

1．认真阅读双迹示波器的工作原理，弄清示波器面板上各主要控制件的作用及不同输入情况下的使用方法。

2．弄清信号发生器各部分的使用方法。

【实验报告要求】

1．用示波器观察信号波形时，要达到波形清晰，亮度适中，波形稳定、改变波形个数，改变波形幅度，应调节哪些旋钮？填入表 6 - 1。

表 6 - 1　示波器调节

波形要求	波形清晰	亮度适中	波形稳定	改变波形个数	改变波形幅度
调节旋钮					

2．用示波器测量给定正弦波、方波、锯齿波信号的幅值和频率，把测量值和信号发生器指示值列表相比较。填入表 6 - 2。

表 6 - 2　波形幅值和频率测量

波形	幅度 U_{P-P}	频率 f 测量值	指示值 f

3．用李萨如图形法测出频率 f_y（f_y 分别为 $1 \times f_x$、$2 \times f_x$ 和 $3 \times f_x$）和信号发生器指示值列表比较。填入表 6 - 3。

表 6 – 3　李萨如图形

波形	 ◯	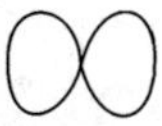 ◯◯	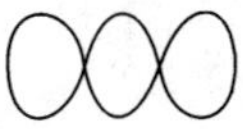 ◯◯◯
测量值 $f_y = f_x \cdot \dfrac{N_x}{N_y}$			
指示值 f_y			

【实验仪器】

1. 双迹示波器（CS – 4125 型）1 台。
2. XD – 82 型多用途信号发生器 1 台。
3. J 9280A 型多用教学仪 1 台。
4. DF1641A 宽频带函数波发生器 1 台。
5. GS1641A 型宽频带函数波发生器 1 台。

6　Basic Electronic Instruments

Objectives

1. Understanding structure and working principle of dual-trace oscilloscope.
2. Learn to operate basic electronic instruments correctly.

Principle and Method

1. Basic electronic instruments meters

During analog circuit lab work we may meet some of the basic instruments and meters, their functions and relationships can be shown in Figure 6 − 1.

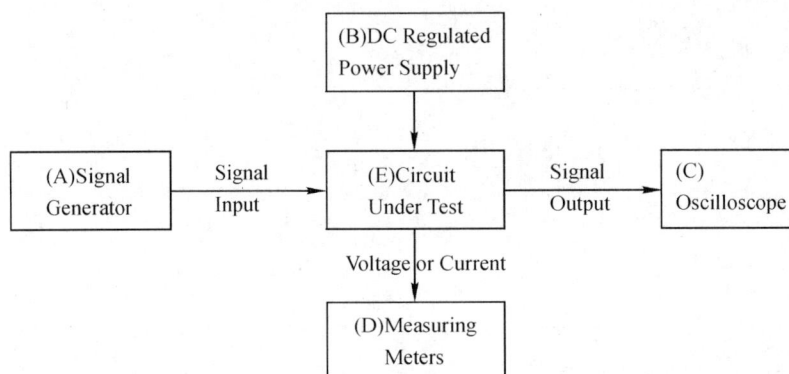

Figure 6 − 1　Relation among analog circuit, instruments and meters

(1) Signal generator: an instrument can produce electric signals including sine wave, triangle wave and rectangle wave, of which the voltage and frequency are all adjustable. The output is indicated digitally or by a pointer. Waveform can be selected according to the circuit under test.

(2) DC regulated power supply: it can supply power for the circuit under test, and usually its output is in a form of voltage, such as: 5 ~ 6V, ±12V, etc.

(3) Oscilloscope: an instrument can observe the output signals of the circuit under test. An oscilloscope can show the voltage and waveform of a signal and can measure frequency, period and other parameters.

(4) Measuring meters: usually this kind of meters including Ohmer, voltmeter, ammeter, frequency meter and so on, for instance, transistorized millivoltmeter, milliammeter, pointer type multimeter, digital multimeter, frequency meter.

(5) Circuit under test: it is studied by instruments or meters to get measuring data and observed. The circuit function should be carefully analyzed.

2. Dual-trace oscilloscope

Dual-trace oscilloscope can observe and measure two signals in a same time. It can not only show the two signals on the screen to be compared, but also can show the amplified difference or addition of two signals. When it is used as a single trace oscilloscope, there is only one channel working.

The characteristic feature of a dual-trace oscilloscope comparing to single-trace oscilloscope is that it can show two traces or signals on the screen in same time. There are two ways to do so: one is double-beam oscilloscope and another is carried out by electronic switch. The CS – 4125 oscilloscope does make use of electronic switch to show two figures on its screen.

Figure 6 – 2 is a diagram of electronic switch. Its working principle is as following: the signals from channel Y_A and Y_B are coupled alternately through electronic switch S, the alternation frequency of the electronic switch must large enough to make your eyes see the two pictures from different channels at a same time.

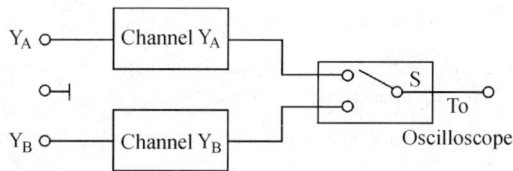

Figure 6 – 2

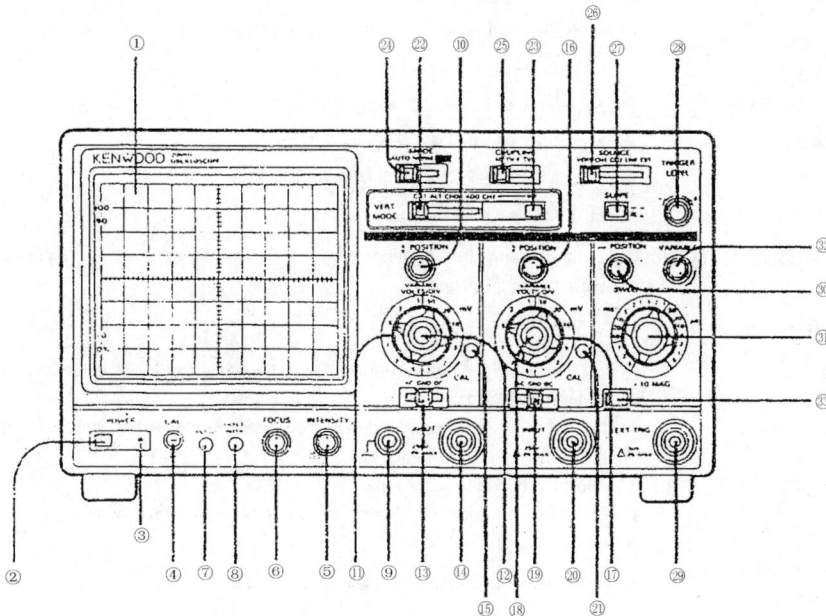

Figure 6 – 3

Figure 6 – 3 is a picture of front panel of CS – 4125 dual-trace oscilloscope. The functions of all the knobs and keys are as following.

No.	Knob Name	Function
1	INTENSITY	Adjust the brightness of the figure on the screen
2	ASTIG	Working with the FOCUS knob, adjust the definition of the figure on the screen
3	FOCUS	Adjust the definition of the figure on the screen
4	TRACE ROTA	Adjust the figure parallel with the horizontal scale
5	CAL	Provide a square wave of amplitude of 1 div (V_{P-P}) and frequency of 1kHz
6	POWER INDICATOR	Light when power on
7	POWER ON/OFF	Switch for power on/off
8	CH1 POSITION	Adjust the vertical position of the figure from channel 1
9	CH2 POSITION	Adjust the vertical position of the figure from channel 2
10	VERTICAL MODE	CH1 or CH2: display the figure from channel 1 or 2 separately ALT: alternatively display figures from the two channels CHOP: chopping display, which can be applied when the scan speed is quit small ADD: Display the sum or difference of the signals from the two channels
11	CH1 VOLTS/DIV	Vertical deflection sensitivity of CH1
12	CH2 VOLTS/DIV	Vertical deflection sensitivity of CH2
13	CH1 VARIABLE	Continuously adjust the vertical sensitivity, at the standard position of clockwise rotation end, can measure signal amplitude
14	CH2 VARIABLE	Continuously adjust the vertical sensitivity, at the standard position of clockwise rotation end, can measure signal amplitude
15	CH 1 AC – GN – DC	Ways of coupling of signal feeding in on CH 1
16	CH 2 AC – GN – DC	Ways of coupling of signal feeding in on CH 2
17	CH 1 INPUT	Input socket for CH 1
18	CH 2 INPUT	Input socket for CH 2
19	GND	The grounding end connected with the oscilloscope shell
20	EXT INPUT	Input socket for external trigger signal
21	INT TRIG SOURCE	Trigger select: CH 1, CH 2 or alternatively
22	TRIG SOURCE	Trigger source select: INT, EXT or LINE
23	SLOPE	Select type of trigger scan: rise edge or fall edge
24	LEVEL	Adjust the trigger scan voltage
25	SWEEP VARIABLE TIME)	Continuously adjust the scanning speed, at the standard position of clockwise rotation end, can measure signal period
26	SWEEP TIME/DIV	Adjust the scanning speed
27	TRIG MODE	NORM: no input no display on the screen; inputting signal can be displayed stably under the adjustment of LEVEL AUTO: no input just display light trace on the screen; inputting signal can be displayed stably under the adjustment of LEVEL TV: Just to show TV signal P – P AUTO: no input no display on the screen; inputting signal can be displayed stably without adjustment of LEVEL
28	TRIG'D	When triggering scanning, the indication light on
29	POSITION PULL × 10	Adjust the horizontal position of the light trace, scanning speed may be enlarged for 10 times when this knob is pulled out

3. Observation of waveform

（1）Adjust the knobs or keys as following

Name of knob	Position	Name of knob	Position
INTEN	Middle	TRIG MODE	AUTO
FOCUS	Middle	TIME/DIV	0.5ms
POSITION CH1 or CH2	Middle	SLOPE	+
VERTICAL MODE	CHOP	TRIG SOURCE	INT
VOLTS / DIV	1V	INT TRIG SOURCE	CH1
VARIABLE	Standard	AC – GN – DC	AC

（2）Press the power button and the power indicator light on. After a while of warm-up, there may appear a light trace on the screen. Adjust the knobs such as INTEN, FOCUS, ASTIG, POSITION, TRACE ROTA for necessary to display a clear and focused light trace on the screen parallel with horizontal scale.

（3）Select VERTICAL MODE

When you are going to observe one signal, put "MODE" key at CH1 or CH2, and the signal can be feed-in from the selected channel. If you want to observe the two signals from the two channels at same time, put the "MODE" key at "ALT". Now the two signals can be displayed on the screen alternatively. The alternative displaying frequency is controlled by scanning period automatically. If the scanning frequency is quite low, the displaying will be twinkle, and the "MODE" key should be put to "CHOP". When the "MODE" key is at "ADD", we can observe the sum of the two signals, but the amplifying set of the two channels should be same. And attention that when the CH2 "POSITION" is at normal position the displaying is CH1 + CH2; when the CH2 "POSITION" is pulled out （PULLINVERT）, the displaying is CH1 – CH2.

（4）Choice of scanning speed.

Scanning range is from $0.2\ \mu s/DIV$ to $0.5S/DIV$, divided into 20 steps by carries of 2 or 5. The fine turning knob can adjust at least 2.5 times continuously. And the clockwise end of the fine turning knob is a standard position, when rotate the fine turning knob at this position, we can get the time parameters of the signals under test by the reading of horizontal distance of the signal and scanning speed. If we want to observe the fine structure of a certain signal, we can expand the signal by 10 times horizontally when pull out the key "POSITION".

4. Measure the AC voltage

Input a sine wave from CH1 or CH2, adjust scanning speed to get a stable display on the screen. Rotate the fine turning knob to the standard position, and adjust the vertical deflection knob to get a suitable figure on the screen. Record the height value（div）of the figure and the V/div value, and the data such as swing value U_{P-P}, voltage amplitude U_m and effective value U can be calculated. (In Figure 6 – 4, $H = 7$.)

The vertical distance between the peak and valley on the waveform is 7div, and the vertical sensitivity is 0.2V/div (that 0.2V per lattice.). So U_{P-P} and effective value U will be:

$$U_{P-P} = 7 \times 0.2 = 1.4V \quad \text{and} \quad U = U_{P-P}/2\sqrt{2} = 1.4V/2.8 = 0.5V.$$

$$U_m = \frac{1}{2} U_{P-P} = 0.7V, \quad U = \frac{U_m}{\sqrt{2}}$$

The horizontal distance between two nearby valley points is 2 lattice, and the sweep speed is 0.5ms/div. So the period of the signal is:

$$T = 2 \times 0.5ms = 1.0ms$$

The frequency is:

$$f = \frac{1}{T} = \frac{1}{10 \times 10^{-3}} \text{ Hz} = 1000\text{Hz} = 1.0\text{kHz}$$

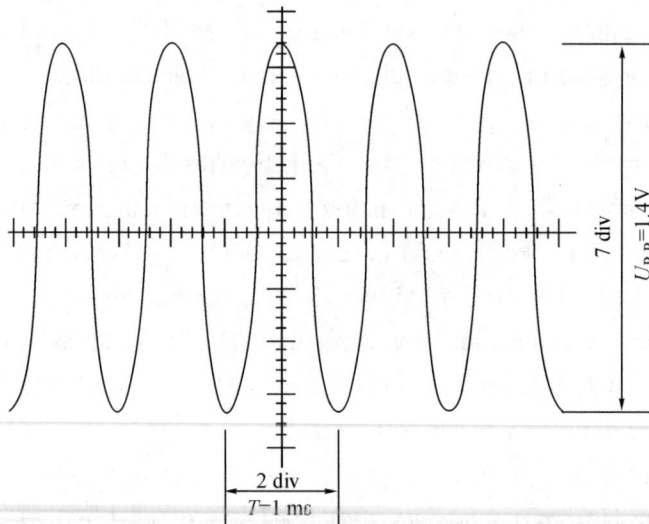

2 div
$T=1$ ms

Figure 6 – 4   A voltage waveform of 0.5V、1kHz

5. Measure frequency

(1) Direct measurement

Separately, input signals of sine wave, square wave and saw-tooth wave into CH 1 or CH 2. Rotate the fine turning knob to the standard position and turn the sweep speed to display 2 waveform of sine wave, 3 square wave and 4 saw-tooth wave, record their waveform width L (in div) and the corresponding value of t/div. Calculate the period T and frequency f:

$$T = L\text{div} \times t/\text{div}, \quad f = 1/T$$

（2）Lissajou figure

Low AC frequency signals are suitable for Lissajou figure method. The process is input two sine signals into x-axis and y-axis, one with known frequency to x-axis and the other for determining frequency to y-axis, the figure displaying on the screen is called Lissajou figure. There are different types of Lissajou figures corresponding to different frequency ratios and initial phases. When the frequency ratio is simple integral the Lissajou figure is a simple clear and closed figure. Three cases of frequency ratio f_y/f_x of 1, 2 and 3 are shown in Figure 6 – 6. Now, let N_x be the number of tangent points by a horizontal line with the Lissajou figure and N_y the points by vertical line. The ratio N_x/N_y will be the reciprocal of the frequency ratio, thus the unknown frequency will be :

$$f_y = f_x \cdot \frac{N_x}{N_y}$$

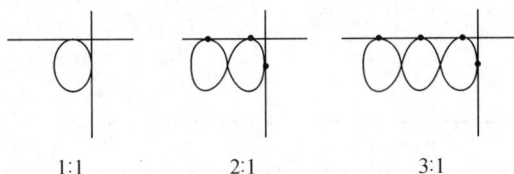

1:1　　　　　　2:1　　　　　　3:1

Figure 6 – 5

When you want to measure frequency by Lissajou figure, you should put the "MODE" at the position of "X-Y ON" first. The sine wave of 50Hz from signal generator is input into x-axis as a standard signal, a sine wave with unknown frequency is input into y-axis. Adjust the frequency of the signal on y-axis and the knobs on oscilloscope panel to get a stable Lissajou figure on screen. Write down the figure and calculate frequency. Compare your result with the frequency produced by the generator.

Preparation Requirement

1. Read the working principle of the dual-trace oscilloscope and understand the functions of the knobs and keys on the panel.

2. Understand the operation of signal generator.

Report Requirement

1. The figure on the screen should be clear and stable with suitable brightness, and the waveform and amplitude should be adjustable. Which knobs should have been adjusted? Fill the Table 6 – 1.

Table 6 – 1   Adjustment of oscilloscope

Requirement for waveform	Clear	Bright	Stable	Change Number	Change Amplitude
Knob					

2. Measure the frequency and amplitude of sine wave, square wave and saw-tooth wave; compare your results with the indication values on the generator. Fill Table 6 – 2.

Table 6 – 2   **Measure Amplitude and Frequency**

Waveform	U_{P-P}	f-calculated	f-indicated

3. Measure frequency by Lissajou figure, let f_y be $1 \times f_x$, $2 \times f_x$ and $3 \times f_x$; compare your results with the indication values on the generator. Fill Table 6 – 3.

Table 6 – 3   **Lissajou Figure**

Waveform			
$f_y = f_x \cdot \dfrac{N_x}{N_y}$			
f_y – indicated			

Equipments

1. Dual-trace oscilloscope, type CS – 4125.
2. Signal generator, type XD – 82.
3. Teaching device, type J9280A.
4. Broad band function generator, type DF1641A.
5. Broad band function generator, type GS1641A.

实验七　RC 放大电路的调试和研究（综合性实验）

【实验目的】

1. 综合研究放大电路参数变化对放大器性能的影响。
2. 学会放大电路的常规调试方法。
3. 测定放大电路的电压放大倍数 A_u，总结出参数变化对 A_u 的影响。

【实验原理与方法】

图 7-1 为分压式偏置电流负反馈带 RC 耦合的放大电路图。

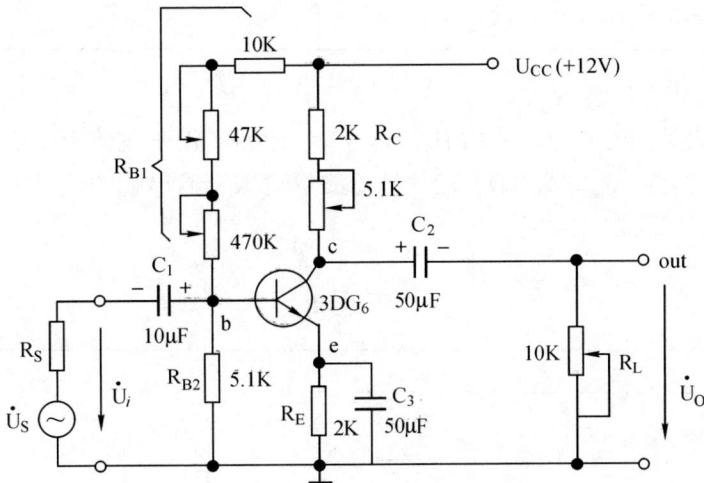

图 7-1

　　分压式偏置电路，是交流放大器中最常用的一种基本电路，这种电路具有工作稳定的优点。它的电压放大倍数

$$A_u = -\beta \frac{R'_L}{r_{be}}$$

　　式中负号表示输出电压与输入电压反相，$R'_L = R_C /\!/ R_L$，因此，电路的静态工作点 Q 参数，电阻 R_C 或 R_L 的变化，都会对放大倍数 A_u 产生影响。

【实验内容和步骤】

1. 电路的分析与计算

　　如图 7-1 所示电路，设 3DG6 的 $\beta = 50$，$R_{B2} = 5.1\text{k}\Omega$，$R_{B1} = 10\text{k}\Omega$，$R_C = 2\text{k}\Omega$，$R_L = 4.7\text{k}\Omega$，列式计算 Q 点（$I_B$、$I_C$、$U_{CE}$）、$r_{be}$、$A_u$、$r_i$ 和 r。

2．认识电路

对照图 7-1 电路，认识实验板上所拼接的电路，找出相应的元件及电路的输入端和输出端。

3．调整放大电路的静态工作点

静态工作点 Q 是否合适，对放大电路的性能和输出波形都有影响。调节 R_{B1}，即可调节 Q 点（I_B、I_C、U_{CE}）之值。实验中，一般避免断开集电极来测 I_C，所以采用测量电压 U_{R_C}，然后算出 I_C 的方法。

$$I_C = \frac{U_{R_C}}{R_C}, \quad I_B = \frac{I_C}{\beta} \quad (\beta = 50)$$

取 $R_C = 2k\Omega$，$R_L = 4.7k\Omega$，调节可调电阻 R_{B1}，使 $U_{CE} = 6V$，用数字万用表 DC 档测量电压，数据记于下表：

U_{CC} (V)	U_B (V)	U_{BE} (V)	U_E (V)	U_{CE} (V)	U_{R_C} (V)	计算 I_C (mA)	计算 I_B (mA)

4．测量电压放大倍数 A_u

从信号发生器中输出频率 $f = 10^3Hz$、幅度 $U_i = 20mV$ 正弦波信号，加到放大电路的输入端，保持 $R_C = 2k\Omega$，$R_L = 4.7k\Omega$，用示波器测量输入、输出端的信号电压，计算电压放大倍数：

$$A_u = -\frac{U_o}{U_i}$$

U_i (mV)	U_o (mV)	A_u

5．研究电路参数变化对放大倍数 A_u 的影响

（1）负载电阻 R_L 变化对 A_u 的影响

电路其他参数不变，输入电压 $f = 10^3Hz$，$U_i = 20mV$，改变 R_L 的值，测出不同 R_L 时的 U_o，并计算 A_u。

R_L (kΩ)	0.51	1.0	3.0	4.7	R_{Lmax}
U_i (mV)					
U_o (mV)					
A_u					

结论：

（2）集电极电阻 R_C 变化对 A_u 的影响

电路其他参数不变，输入电压 $f = 10^3 \text{Hz}$，$U_i = 20\text{mV}$，R_L 调至最大，改变 R_C 值，测出不同 R_C 时的 U_o，并计算 A_u。

R_C (kΩ)	2.0	4.7	6.2	7.0	8.0	9.0	10.0
U_i (mV)							
U_o (mV)							
A_u							

结论：

(3) 集电极电流 I_C 的变化对 A_u 的影响

保持 $R_C = 2.0\text{k}\Omega$，$R_L = 4.7\text{k}\Omega$，调节可变电阻 R_{B1}，使 U_{R_C} 变化，从而改变 I_C（$I_C = U_{R_C}/R_C$），测出不同 I_C 值时的 U_i、U_o 并计算 A_u。（注：当 U_o 出现失真时，应适当减小 U_i 值）

I_C (mA)	0.6	0.8	1.0	1.2
U_i (mV)				
U_o (mV)				
A_u				

结论：

6. 观察工作点不合适时引起的波形失真

取电路参数 $R_C = 2\text{k}\Omega$，$R_L = 4.7\text{k}\Omega$，调节上偏电阻 R_{B1}，并逐渐增大输入信号，使输出信号出现削波，分别观察出现何种类型的失真，并画下失真的波形。

R_{B1}	U_o 失真波形	失真类型（饱和/截止）
偏大时		
偏小时		

7. 放大器最佳工作点与最大不失真输出电压的观察

取电路参数 $R_C = 2\text{k}\Omega$，$R_L = 4.7\text{k}\Omega$，调节上偏电阻 R_{B1}，使得当输入电压逐渐增大时，输出波形正负向同时出现削波，此即表示放大器的静态工作点已选在动态特性曲线的中点，称为最佳工作点。记录此时的 I_{CQ}，U_{CEQ}，最大不失真 U_i 和 U_o，可以看到，此时放大器的动态范围最大。

I_{CQ} (mA)	U_{CEQ} (V)	U_i (mV)	U_o (mV)

【**实验预习要求**】

1. 了解放大器的基本原理和测试方法。

2．计算实验内容中的 Q、A_u、r_i、r_o 值。

【实验报告要求】

1．列出并整理实验数据与波形。

2．分析电路参数 I_C、R_C、R_L 变化对放大器性能的影响，为什么？写出放大倍数随参数改变而变化的结论？

3．完成预习要求中的计算。

4．回答如下问题：

（1）放大电路为什么要设置静态工作点？何为最佳静态工作点？

（2）放大电路带上负载后电压放大倍数如何变化？

（3）分析输出波形失真原因，提出解决办法。

U_o 输出波形				
状态	正常放大			
解决方法	R_{B1} 值大小合适			

【实验仪器】

1．双迹示波器 1 台。

2．信号发生器 1 台。

3．直流稳压电源 1 台。

4．数字万用表 1 块。

5．放大器线路实验板 1 块。

7 Study on RC Amplifying Circuit (Integration Experiment)

Objectives

1. Study the influence of circuit parameters on the performance of amplifying system.
2. Learn the normal debug method for amplifying circuit.
3. Measure the voltage gain A_u and explain the influence of parameters on A_u.

Principle and Method

Figure 7 – 1 shows a voltage-divider current negative feedback with RC coupling amplifying circuit.

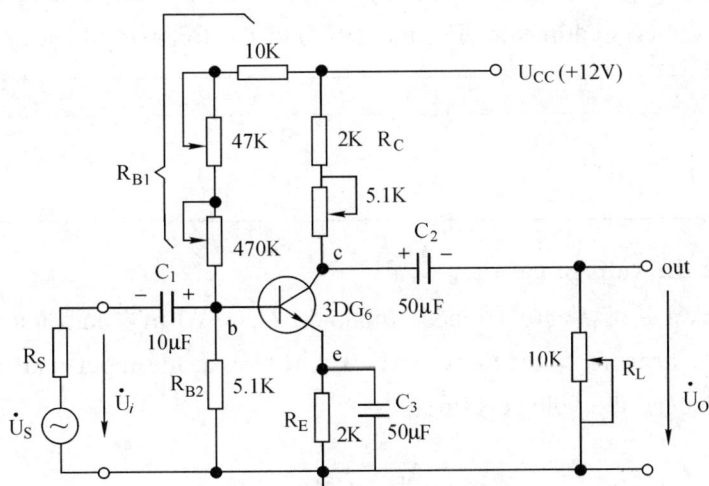

Figure 7 – 1

Voltage-divider biasing circuit is one of the most common AC amplifying circuit, which has a characteristic of steady working. The voltage gain is:

$$A_u = -\beta \frac{R'_L}{r_{be}}$$

The negative sign indicates that output voltage is out of phase with the input voltage. And $R'_L = R_C /\!/ R_L$. Thus, the quiescent point Q, the resistance R_C and R_L all will affect the voltage gain A_u.

Experiment

1. Analyze and calculate

For the circuit in Figure $7-1$, for the transistor 3DG6 $\beta = 50$, $R_{B2} = 5.1\text{k}\Omega$, $R_{B1} = 10\text{k}\Omega$, $R_C = 2\text{k}\Omega$, $R_L = 4.7\text{k}\Omega$. Calculate the quiescent point Q (I_B, I_C and U_{CE}), r_{be}, A_u, r_i and r_o.

2. Understand the actual circuit

Understand the actual circuit on the experiment board and compare it with Figure $7-1$. Recognize all the electric devices and the input and output ends.

3. Debug the quiescent point

The quiescent point Q has effects on the performance of an amplifying circuit and the output waveform. Adjust R_{B1} to change the values I_B, I_C and U_{CE} of point Q. Actually, we can measure the voltage U_{R_C} to get I_C, avoiding break the collector:

$$I_C = \frac{U_{R_C}}{R_C}, \quad I_B = \frac{I_C}{\beta} \quad (\beta = 50)$$

Let $R_C = 2\text{k}\Omega$, $R_L = 4.7\text{k}\Omega$. Adjust R_{B1} to $U_{CE} = 6\text{V}$, which can be measured by DC voltage step of a digital multimeter. Fill the data into the following table.

U_{CC} (V)	U_B (V)	U_{BE} (V)	U_E (V)	U_{CE} (V)	U_{R_C} (V)	I_C (mA)	I_B (mA)

4. Measure the voltage gain A_u

Input a sine wave of $f = 10^3\text{Hz}$ and amplitude $U_i = 20$ mV onto the input end of the amplifying circuit, keep $R_C = 2\text{k}\Omega$, $R_L = 4.7\text{k}\Omega$. Observe the input and output signals on oscilloscope. Calculate the voltage gain:

$$A_u = -\frac{U_o}{U_i}$$

U_i (mV)	U_o (mV)	A_u

5. Effect of circuit parameters on A_u

(1) Effect of varying of R_L on A_u

Keep the circuit parameter unchanged. The frequency of input signal $f = 10^3\text{Hz}$, amplitude $U_i = 20\text{mV}$. Change R_L for different values and measure corresponding U_o, calculate A_u.

R_L (kΩ)	0.51	1.0	3.0	4.7	R_{Lmax}
U_i (mV)					
U_o (mV)					
A_u					

Conclusion:

(2) Effect of collector resistance R_L on A_u

Keep the circuit parameter unchanged. The frequency of input signal $f = 10^3$Hz, amplitude $U_i = 20$mV. Let R_L be the maximum value. Change R_C for different values and measure corresponding U_o, calculate A_u.

R_C (kΩ)	2.0	4.7	6.2	7.0	8.0	9.0	10.0
U_i (mV)							
U_o (mV)							
A_u							

Conclusion:

(3) Effect of collector current I_C on A_u

Keep $R_C = 2.0$kΩ, $R_L = 4.7$kΩ. Adjust R_{B1} to change U_{R_C}, and thus I_C ($I_C = U_{R_C}/R_C$) changes. Measure U_i and U_o under different I_C, and calculate A_u. (Note: if distortion happens to U_o, adjust U_i smaller.)

I_C (mA)	0.6	0.8	1.0	1.2
U_i (mV)				
U_o (mV)				
A_u				

Conclusion:

6. Observe wave distortion causing by unsuitable quiescent point Q

Let $R_C = 2$kΩ and $R_L = 4.7$kΩ. Adjust biasing resistance R_{B1}. Increase the input signal to make the output clipping distortion. Estimate the type of distortion and draw the distortion waveform.

R_{B1}	Distortion Waveform of U_o	Distortion Type (Saturation/Cut-off)
Too Large		
Too Small		

7. Best operation point and most undistorted output voltage

Keep $R_C = 2$kΩ, $R_L = 4.7$kΩ. Adjust biasing resistance R_{B1} to a certain value that when increase input voltage clipping distortions appears to the top and bottom part of output waveform. Therefore, the quiescent point is at the center of the dynamic performance curve, and it is the best operation point. Record I_{CQ}, U_{CEQ}, most undistorted U_i and U_o. And in this circumstance the dynamic working range is the largest.

I_{cq} (mA)	U_{CEQ} (V)	U_i (mV)	U_o (mV)

Preparation Requirement

1. Understand the basic principle test method for a amplifying system.
2. Calculate the values of Q, A_u, r_i and r_o.

Report Requirement

1. List your data and waveforms.
2. Analyze effect of the varying of parameter I_C, R_C and R_L on the performance of amplifying system. What affect the voltage gain?
3. Calculate the values put forward in last section.
4. Answer the following questions:

(A) It is always setup a quiescent point in amplifying circuit, why? What's the best quiescent point?

(B) How does the voltage gain change after feed a load?

(C) Analyze the reasons of wave distortion, and discuss solving methods.

Output Waveform U_o				
State	Normal			
Solving Method	Suitable R_{B1}			

Equipments

1. Dual-trace oscilloscope.
2. Signal generator.
3. DC Power supply.
4. Digital multimeter.
5. RC amplifying circuit experiment board.

实验八 射极输出器的调试和研究

【实验目的】

1. 学会测量射极输出器的输入电阻、输出电阻及其电压跟随范围。
2. 掌握射极输出器的特点。

【实验原理与方法】

射极输出器是负反馈放大器的特例，它的电路图如图 8 – 1 所示。

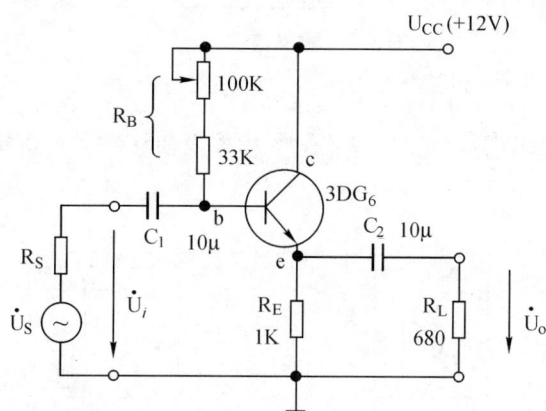

图 8 – 1

射极输出器的特点是：电压放大倍数小于 1 而近于 1，输出电压与输入电压同相，输入电阻高，输出电阻低。

射极输出器的电压放大倍数 A_u：

$$A_u = \frac{\beta R'_L}{r_{be} + \beta R'_L}$$

式中

$$R'_L = \frac{R_E R_L}{R_E + R_L}$$

输入电阻 r_i

$$r_i = r_{be} + (1 + \beta) R'_L$$

输出电阻 r_o

$$r_o = \frac{R'_S + r_{be}}{\beta}$$

式中

$$R'_S = \frac{R_B R_S}{R_B + R_S}$$

输入电阻和输出电阻的测量方法：

（1）输入电阻的测量：

在输出端接有 R_L 的情况下，将一个已知阻值的附加电阻 R 串入输入电路，如图 8-2 所示，当外加输入信号 U_s 后，测出 U_s 和 U_i 的数值，便可计算出输入电阻 r_i。

$$r_i = U_i / I_i$$

而

$$I_i = （U_s - U_i）/R$$

则测出 U_i、U_s，可得输入电阻

$$r_i = R\left(\frac{U_i}{U_s - U_i}\right)$$

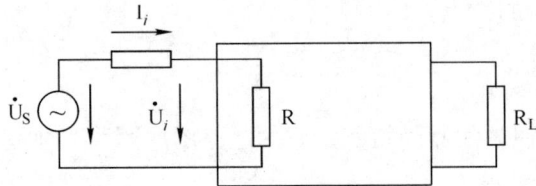

图 8-2

（2）输出电阻的测量：

从图 8-3 可知，测出当负载电阻 R_L 在断开和接通这两种情况的输出电压，便可计算出输出电阻 r。

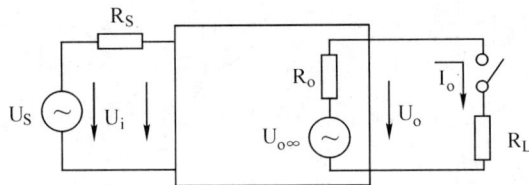

图 8-3

设输出端开路电压为 $U_{o\infty}$，而 R_L 接通时的输出电压为 U_o，则

$$r_o = \frac{U_{o\infty} - U_o}{I_o} = \frac{U_{o\infty} - U_o}{U_o} \cdot R_L$$

【实验内容和步骤】

1. 电路的分析计算

如图 8-1 所示电路，若 3DG6 的 $\beta = 50$，$R_s = 50\Omega$，求 Q（I_B、I_E、U_{CE}）、A_u、r_i 和 r。

2. 认识电路

对照图 8-2 电路，认识实验板上电路，找出相应的元件及电路的输入端和输出端。

3. 测量静态工作点

调节 R_B，使 $U_{CE} = 6V$，电路不输入信号。用数字万用表测量下表所列各电压值，并计算 I_B、I_E，数据记录于下表 $[I_E = U_E/R_E，I_B = I_E/（1+\beta）]$。

U_{CC}（V）	U_B（V）	U_{BE}（V）	U_E（V）	U_{CE}	计算 I_E（mA）	计算 I_B（mA）

4．测量电压放大倍数 A_u

多用途信号发生器输出 $f = 10^3$ Hz、$U_S = 1$V 的正弦波信号加到被测电路的输入端，用示波器测量输入、输出的信号电压值。改变负载电阻 R_L 的数值，比较电压放大倍数是否变化，画出电压放大倍数随 R_L 的变化曲线。

负载电阻 R_L（Ω）	100	680	5.1K	开路
输出电压 U_o（V）				
电压放大倍数 A_u				

5．测量输入电阻 r_i

（1）按图 8-4 连接电路，$f = 10^3$ Hz、$U_S = 1$V 的正弦波输入信号 U_s 经可变电阻 R_i 与被测电路的输入端连接。

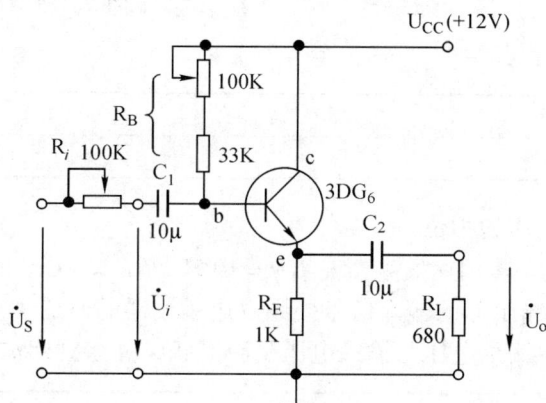

图 8-4

（2）用示波器测量基极的输入电压 U_i，调节 R_i 由 0 值逐渐增大，直到 U_i 下降为原来数值的一半，这时 R_i 的阻值就等于射极输出器的输入电阻 r_i。

$$r_i = R_i \frac{U_i}{U_s - U_i} = R_i \left(\frac{U_s/2}{U_s - U_s/2} \right) = R_i$$

（3）测量结果记于下表。

U_i（V）	1	0.5
R_i（Ω）	0	

6．测量输出电阻 r_0

（1）按图 8-5 连接电路，多用信号发生器输出 $f = 10^3$ Hz、$U_S = 30$mV 的正弦波信号 U_s 加到被测电路的输入端。

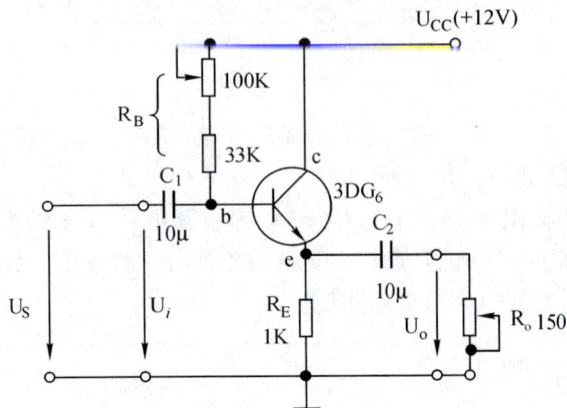

图 8 – 5

（2）先不接 R_o，测得的输出电压为 $U_{o∞}$，再接入可变电阻 R_o，调节使输出电压 $U_o = \frac{1}{2} U_{o∞}$，此时的 R_o 阻值就等于射极输出器的输出电阻 r_o。

$$r_o = \frac{U_{o∞} - U_o}{U_o} R_o = \frac{U_{o∞} - U_{o∞}/2}{U_{o∞}/2} R_o = R_o$$

（3）测量结果记入下表

U_o (mV)	$U_{o∞}$	$U_{o∞}/2$
R_o (Ω)	开　路	

7．观察输入输出电压的相位

电路连接同图 8 – 1，多用途发生器输出 $f = 10^3\text{Hz}$、$U_S = 30\text{mV}$ 的正弦波信号加到被测电路的输入端。将射极输出器的输入信号接到双迹示波器的通道 1（CH – 1），输出信号接到通道 2（CH – 2）观察输出电压与输入电压是否同相，并将观察波形记于下表。

U_i 的波形	
U_0 的波形	

8．测量电压跟随范围

（1）电路同上。逐渐增大输入信号电压，观察示波器上的输出电压波形，测出最大不失真的输出电压，此电压又称最大跟随电压。

记下最大不失真输出电压 $U_o = $ _____ （V）

（2）调节 R_B，使得当输入电压逐渐加大时，输出波形正负向同时出现削波，此时即表示射极输出器的静态工作点已选在交流负载线的中点。然后测出最大不失真输入电压的峰－峰值 U_{iP-P}，此值叫做射极输出器的最大电压跟随范围。

最大电压跟随范围 $U_{iP-P} = $ _____ （V）

【实验预习要求】

1．深入了解负反馈对放大器性能的影响。

2．完成实验内容 1 的计算。

3．掌握射极输出器的工作原理和特点。

4．明确实验内容和方法步骤。

【实验报告要求】

1．画出实验线路，标明元件数值。

2．将实验结果整理列表，并与计算值比较。

3．根据实验现象和数据分析，总结射极输出器有哪些优点。

【实验仪器】

1．双迹示波器 1 台。

2．信号发生器 1 台。

3．直流稳压电源 1 台。

4．数字万用表 1 块。

5．射极输出器线路实验板 1 块。

8　Study on Emitter Follower

Objectives

1. Measure the input and output resistance of an emitter follower and its voltage following range.

2. Master the characteristics of an emitter follower.

Principle and Methods

Emitter follower is a special case of negative feedback amplifier. Its circuit is shown in Figure 8 – 1.

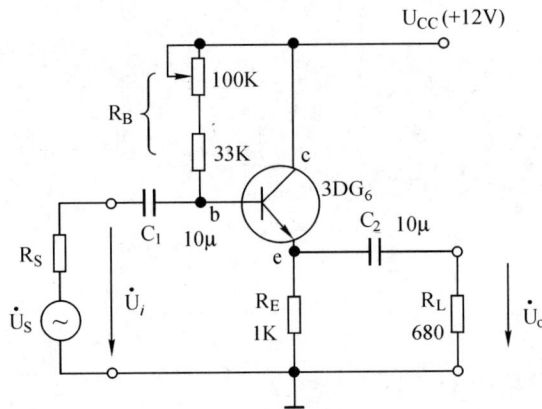

Figure 8 – 1

The characteristics of an emitter follower are: its voltage gain is smaller than 1 but near 1; output voltage is the same with input; has high input resistance and low output resistance.

The voltage gain A_u of an emitter follower is:

$$A_u = \frac{\beta R'_L}{r_{be} + \beta R'_L},$$

where:

$$R'_L = \frac{R_E R_L}{R_E + R_L}$$

The input resistance of an emitter follower r_i is $r_i = r_{be} + (1 + \beta) R'_L$, and the output resistance r_o is $r_o = \dfrac{R'_s + r_{be}}{\beta}$,

where:

$$R'_S = \frac{R_B R_S}{R_B + R_S}.$$

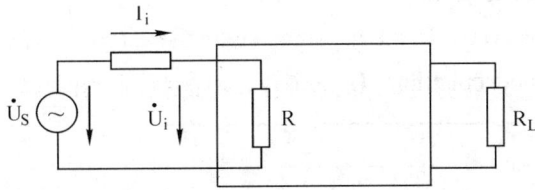

Figure 8 – 2

Methods for measuring input and output resistance:

(1) Input resistance

When resistance R_L is joined at the output end of the circuit, set a resistance R in series into the input part of the circuit, as shown in Figure 8 – 2. Add a signal U_s, measure U_i, and calculate the input resistance r_i:

$$r_i = U_i / I_i$$

where:
$$I_i = (U_s - U_i) / R$$

And input resistance: $r_i = R \left(\dfrac{U_i}{U_s - U_i} \right)$

(2) Output resistance

According to Figure 8 – 3, we can measure the output voltages for the two cases of load resistance R_L joining in the circuit or not, and calculate the output resistance.

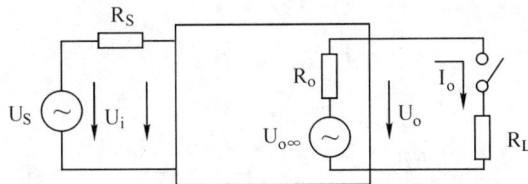

Figure 8 – 3

Let $U_{o\infty}$ be open circuit voltage on the output end, U_o is output voltage with the joining of R_L. The output resistance r_o is:

$$r_o = \frac{U_{o\infty} - U_o}{I_o} = \frac{U_{o\infty} - U_o}{U_o} \cdot R_L$$

Experiment

1. Analyze and calculate

For the circuit in Figure 8 – 1, for the transistor 3DG 6 $\beta = 50$, $R_s = 50\ \Omega$. Calculate the quiescent point Q (I_B, I_E and U_{CE}), A_u, r_i and r_o.

2. Understand the actual circuit

Understand the actual circuit on the experiment board and compare it with Figure 8 – 2. Recognize all the electric devices and the input and output ends.

3. Measure quiescent point

Adjust R_B to $U_{CE} = 6V$. Input no signal into the circuit. Measure all the voltages list in the following table, and calculate I_B and I_E. ($I_E = U_E/R_E$, $I_B = I_E/(1+\beta)$)

U_{CC} (V)	U_B (V)	U_{BE} (V)	U_E (V)	U_{CE}	I_E (mA)	I_B (mA)

4. Measure voltage gain A_u

A sine wave of $f = 10^3$Hz and $U_S = 1V$ from signal generator is send to the input end of the circuit under test. The input and output signals are observed by the oscilloscope. Compare the voltage gains under different values load resistance R_L. Draw the curve of voltage gain A_u and R_L.

Load Resistance R_L (Ω)	100	680	5.1K	Open Circuit
Output Voltage U_o (V)				
Voltage gain A_u				

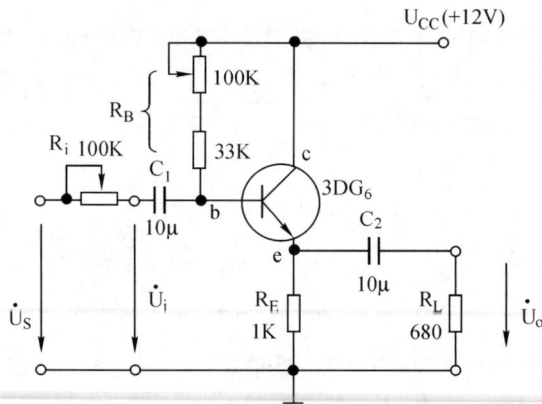

Figure 8 – 4

5. Measure the input resistance r_i

(1) Join the circuit according to Figure 8 – 4. A sine wave of $f = 10^3$Hz and $U_S = 1V$ is input into the circuit under test through variable resistance R_i.

(2) Observe the base input voltage U_i on the oscilloscope. Increase R_i from 0 till the U_i decrease to half of its original value. And now the value of R_i equals the input resistance r_i of the emitter follower.

$$r_i = R_i \frac{U_i}{U_s - U_i} = R_i\left(\frac{U_s/2}{U_s - U_s/2}\right) = R_i$$

(3) Record your measurement:

U_i (V)	1	0.5
R_i (Ω)	0	

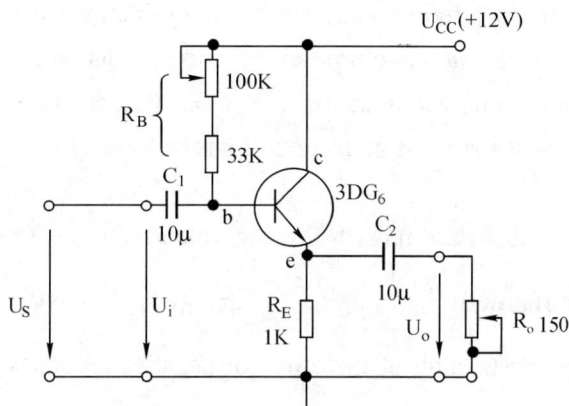

Figure 8 – 5

6. Measure the output resistance r_o

(1) Join the circuit according to Figure 8 – 5. A sine wave of $f = 10^3$Hz and $U_S = $ 30mV from signal generator is input into the circuit under test.

(2) Measure the output voltage $U_{o\infty}$ without R_o. Then join in the variable resistance R_o, and adjust it till the output voltage $U_o = \dfrac{1}{2} U_{o\infty}$. Now, the value of R_o is the output resistance r_o of the emitter follower:

$$r_o = \frac{U_{o\infty} - U_o}{U_o} R_o = \frac{U_{o\infty} - U_{o\infty}/2}{U_{o\infty}/2} R_o = R_o$$

(3) Record your measurement:

U_o (mV)	$U_{o\infty}$	$U_{o\infty}/2$
R_o (Ω)	Open Circuit	

7. Observe the phases of input and output voltage

For the same circuit in Figure 8 – 1, input a sine wave of $f = 10^3$Hz and $U_S = 30$mV to the circuit. Connect the signal from the input end of the circuit to CH1 of the oscilloscope and the signal from output to CH2. Observe their phase of the two signal, and fill the following table.

Waveform of U_i	
Waveform of U_o	

8. Measure the voltage following range of emitter follower

(A) For the same circuit in Figure 8 – 1, gradually increase voltage of the input signal and observe the waveform of output on the oscilloscope to get the most undistorted output voltage. It is also called most following voltage.

Write down your result of the most undistorted output voltage: $U_o =$ _____ (V)

(B) Adjust R_B to some certain value, so that when increase the voltage of input signal the top and bottom parts of output waveform appear clipping distortion. It means that the quiescent point of the emitter follower is at the center of AC performance line. And now the peak-to-peak value U_{iP-P} of the most undistorted input voltage is the most following range of emitter follower.

Write down your result of the most following voltage range: $U_{iP-P} =$ _____ (V)

Preparation Requirement

1. Make a good understanding of the effect of negative feedback on the performance of an amplifier.

2. Calculate the relative quantities in last section.

3. Understand the working principle of emitter follower.

4. Know experiment content and steps.

Report Requirement

1. Draw the circuit figure, and mark all the values on elements.

2. List your measure results with your calculation result.

3. According to your experiment, summarize the characteristic of a emitter follower.

Equipments

1. Two-trace oscilloscope.

2. Signal generator.

3. DC power supply.

4. Digital multimeter.

5. Emitter follower circuit experiment board.

实验九　集成运算放大器在信号运算方面的应用

【实验目的】

1．熟悉集成运算放大器的性能。
2．应用集成运算放大器实现基本运算关系。

【实验原理与方法】

　　线性集成电路实际上就是高放大倍数的直流放大器，外接深度电压负反馈电路后便构成运算放大器。运算放大器可对电信号进行数学运算，它是电子模拟计算机的基本单元，在自动控制系统和量测设备中获得广泛的应用。

　　本实验使用 F004 和 OP07 两种集成运算放大器。前者金属 TO 形封装，后者为塑料双列直插式封装。

　　集成运算放大器 F004（5G23），其主要参数为：开环差模电压放大倍数 $10^4 \sim 10^5$，最大输出电压 ± 10V，输入电阻 50 ~ 200kΩ，输出电阻 4000Ω，共模抑制比 76 ~ 86dB。它的外引线为 2 – 反相端，3 – 同相端，6 – 输出端，7 – 正电源，4 – 负电源，1、8 调零端，5 – 相位补偿端。图 9 – 1 是用集成运放 F004 接成反相比例放大器时外部接线的一种方案。图中 R_W 作调零用，电容 C 用作相位补偿。OP07 功能与 F004 相同，两种集成运算放大器的管脚分布见附录五《集成运算放大器管脚分布示意图》。

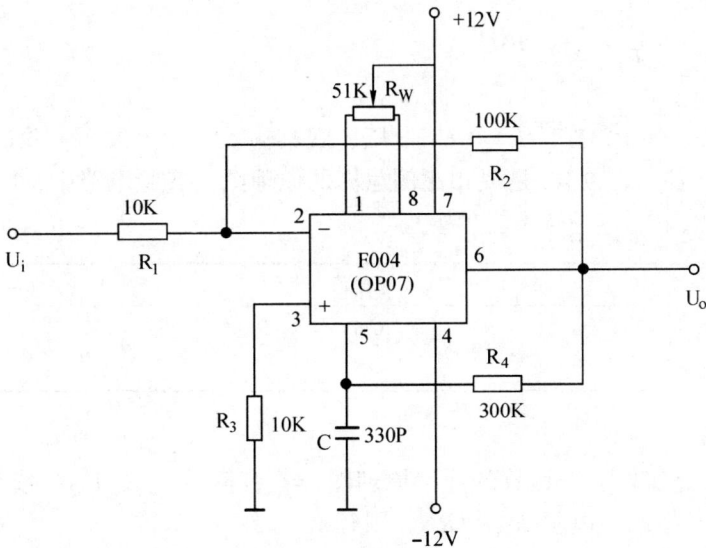

图 9 – 1

【实验内容和步骤】

1. 连接电路

（1）将运算放大器通用实验板连接成图 9 - 1 所示的电路

（2）± 12V 稳压电源连接如图 9 - 2 所示。

（3）注意事项：

①电源电压接反或电源电压突变、输入电压过大时，都会损坏集成运放。

②在改接线路前，必须切断电源和信号源，否则易损坏集成运放。

③不能将集成运放输出端对地短接。

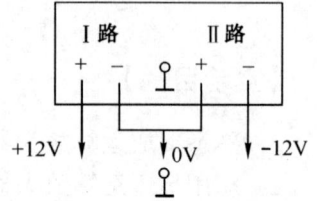

图 9 - 2

2. 调零

将 U_i 反相输入端接地，经检查后，接通电源。调节 R_W 电位器使输出电压 U_O 尽可能精确地指零。

3. 反相比例运算

（1）图 9 - 3 为实现反相比例运算的原理图，设反馈电阻为 R_2，输入端外接电阻 R_1，则有 $U_o/U_i = -R_2/R_1$，改变 R_2 或 R_1 的大小，便可改变其比例大小。

为使两个输入端直流保持平衡，电路中要求 $R_3 = R_1 // R_2$。

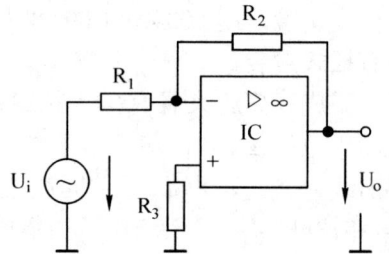

图 9 - 3

（2）电路计算

按图 9 - 2 所示电路，已知 $R_1 = 10\text{k}\Omega$，$R_2 = 100\text{k}\Omega$，R_3 应等于 $R_1 // R_2 = 9.1\text{k}\Omega$，取近似值 10K，则 $U_o = -\dfrac{R_2}{R_1}U_i = -10U_i$。

（3）数据测量

在反相端输入 $f = 10^3\text{Hz}$，$U_i = 0.6\text{V}$ 电压，观察输入、输出波形，验证是否符合 $U_o = -10U_1$ 的关系。（注：图 9 - 3 是反相比例运算的原理图，实验电路中，电源、调零电阻、平衡电阻、补偿电阻电容一样不能少接）。

信号	电压值	$U - t$ 曲线
U_i	0.6V	
U_o		

4. 加法运算

（1）图 9 - 4 是实现加法运算的连接原理图，若有信号 U_{i1}、U_{i2}需按一定比例组合在一起，输入端应外接两个电阻 R_{11} 和 R_{12}，则有

$$U_o = -\left(\frac{R_2}{R_{11}}U_{i1} + \frac{R_2}{R_{12}}U_{i2} \right)$$

实现了加法运算。

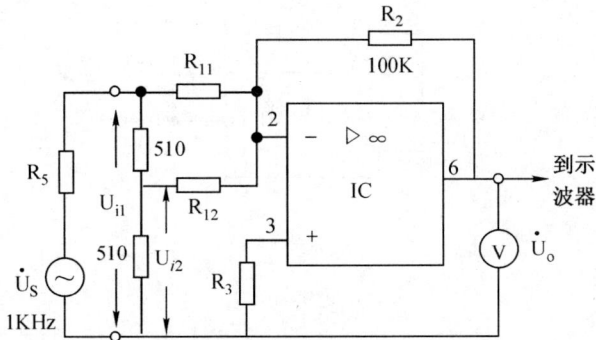

图 9 - 4

（2）设 $R_2 = 100\mathrm{k}\Omega$ 要求 $U_\mathrm{o} = -(10U_{i1} + 10U_{i2})$，为满足上述运算关系计算出：$R_{11} = 10\mathrm{k}\Omega$，$R_{12} = 10\mathrm{k}\Omega$，$R_3 = 5\mathrm{k}\Omega$。

（3）根据计算结果和原理图 9 - 4，在实验板上连接好加法运算电路。

（4）输入 $f = 10^3\mathrm{Hz}$，$U_\mathrm{i} = 0.6\mathrm{V}$ 的信号，经两个串联的 510Ω 电阻分压后加到运算放大器的两个输入端。此时 $U_{i1} = 0.6\mathrm{V}$，$U_{i2} = 0.3\mathrm{V}$。测量方法同前，验证是否符合 $U_\mathrm{o} = -(10U_{i1} + 10U_{i2})$。

信号	电压值	$U - t$ 曲线
U_{i1}	0.6V	
U_{i2}	0.3V	
U_o		

5．同相运算

（1）如输入信号从集成运放的同相端接入，便可构成如图 9 - 5 所示的同相放大器。其输出电压为：

$$U_\mathrm{o} = (1 + R_2/R_1)U_\mathrm{i}$$

（2）设 $R_2 = 100\mathrm{k}\Omega$，要求 $U_\mathrm{o} = 11U_\mathrm{i}$，可计算出 $R_1 = 10\mathrm{k}\Omega$，$R_3 = 10\mathrm{k}\Omega$。

（3）根据原理图 9 - 5 和计算的 R_1，R_3 值，在实验板上连接同相运算电路。

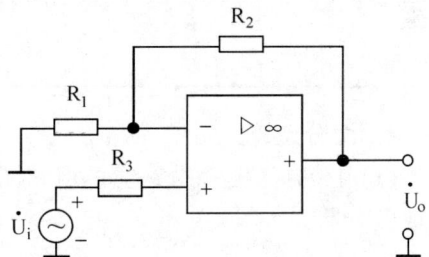

图 9 - 5

（4）输入 $f = 10^3\mathrm{Hz}$，$U_\mathrm{i} = 0.6\mathrm{V}$ 的信号，验证是否符合 $U_\mathrm{o} = 11U_\mathrm{i}$。

信号	电压值	$U - t$ 曲线
U_i	0.6V	
U_o		

6．电压跟随器

（1）图 9 - 6 所示的电路称为电压跟随器，在实验板上按原理图 9 - 6 接成电压跟随器，取 $R_2 = 10\mathrm{K}$，$R_3 = 10\mathrm{K}$。

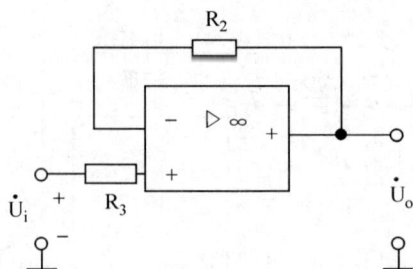

图 9-6

（2）输入 $f = 10^3 Hz$，$U_i = 0.6V$ 的信号，验证电压跟随情况，是否符合 $U_o \cong U_i$。

信号	电压值	$U-t$ 曲线
U_i	0.6V	
U_o		

7．积分运算

（1）图 9-7 是积分运算的电路图，反馈阻抗用电容，输入端外接阻抗用电阻，便构成积分器。

（2）根据图 9-7 电路，在实验板上接成积分运算电路。

当 U_i 为矩形波时，U_o 便是三角波，它是矩形波电压经积分后的结果。

（3）在输入端输入峰-峰值 $U_{iP-P} = 0.6V$ 的方波信号，用示波器观察波形，并将输入、输出的电压波形描绘下来。

信 号	电 压 波 形 $U-t$
U_i	
U_o	

8．微分运算

（1）图 9-8 是微分运算的电路图，反馈阻抗用电阻，而输入端外接阻抗用电容，便构成微分器。

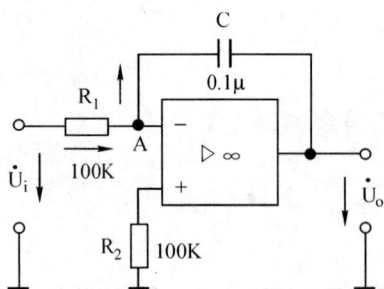

图 9-7

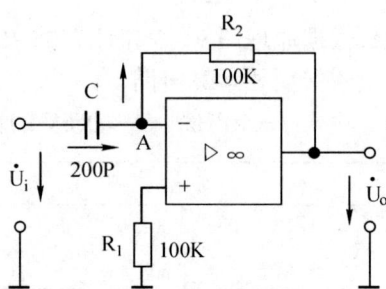

图 9-8

根据图 9-8 电路，在实验板上接成微分运算电路。

当 U_i 为矩形波时，U_o 便为两个正负相同的窄脉冲波，它是矩形电压经微分后的结果。而当 U_i 为三角波时，经微分后，U_o 则为矩形波。

（2）在输入端输入峰–峰值 $U_{iP-P} = 0.6V$ 的方波信号，观察并描绘输入、输出电压的波形。

信　号	电压波形 $U-t$
U_i	
U_o	

9．标度加法器

（1）在测量仪器中，用运算放大器的加法电路，可以作成标度器，图 9–9 所示的标度加法器，就是用正弦波为基准，测量方波周期和宽度的电路。

（2）根据图 9–9 电路，在实验板上接成标度加法器。

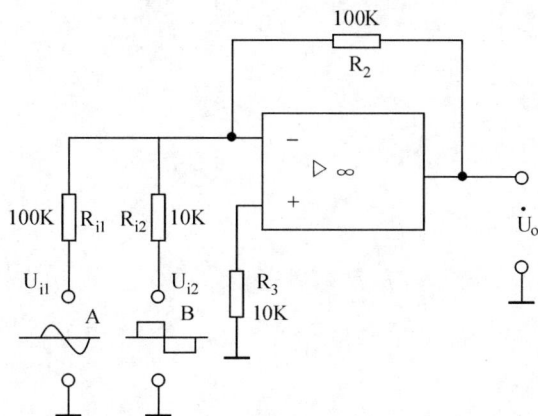

图 9–9

（3）从信号发生器输出 $f = 2000Hz$，$U_{i1P-P} = 10mV$ 的正弦波作为基准信号，输入到标度加法器的 A 端；而从 B 端输入 $f = 100Hz$，$U_{i2P-P} = 600mV$ 的方波作为待测信号。观察 U_{i1}，U_{i2}，U_o 的波形，并测出方波的周期 T。

信　号	电　压　值	$U-t$ 曲线
U_{i1}	10 mV	
U_{i2}	600 mV	
U_o		

测量值方波：$T =$ ＿＿＿＿＿＿ ms　　理论值 $T =$ ＿＿＿＿＿＿ ms

【实验预习要求】

1．了解运算放大器的工作原理和运算关系，熟悉比例器、跟随器、积分器和微分器的电路特点。

2．了解集成运放 F004 和 OP07 外引线排列、名称及其主要技术参数。（参见附录五）

3．明确实验内容及其实施方法。

【实验报告要求】

1．写出反相、比例、同相、跟随和加法运算的实验数据和波形。
2．画出积分、微分和标度加法运算时输入输出电压的波形。
3．写出集成运放的调整、测试的心得体会。

【实验仪器】

1．双迹示波器 1 台。
2．信号发生器 1 台。
3．运算放大器通用实验板 1 块。
4．数字万用表 1 只。

9 Integrated Operational Amplifier: Signal Operation

Objectives

1. Familiar with integrated operational amplifier.
2. Carry out basic operation by integrated amplifier.

Principle and Method

Linear integrated circuit is some kind of DC amplifiers with extremely high gain coefficient. When it is joined with deep negative feedback circuit, the whole circuit is a kind of integrated operational amplifier. Integrated operational amplifier can manipulate electric signals mathematically. It is a basic unit of an analog computer, and is widely adopt in auto control systems and measuring equipments.

There are two types of integrated operational amplifiers in this experiment: F004 and OP07. The former is in metal TO-shape package and the later is dual inline package.

The main parameters of integrated operational amplifier F004 (5G23) are: open-loop differential-mode voltage gain $10^4 \sim 10^5$, most output voltage $\pm 10V$, input resistance $50 \sim 200k\Omega$, output resistance 4000Ω, common-mode rejection ratio $76 \sim 86dB$. Its leg wires are: 2 – reversed phase end; 3 – in phase end; 6 – output end; 7 – positive end of power; 4 – negative end of power; 1 – and 8 – zeroing ends; 5 – phase compensation end. Figure 9 – 1 shows the circuit of an inverse-phase proportional amplifier constructed upon integrated operational amplifier F004. In the circuit, R_W is a zeroing resistance and C is a compensation capacitor. Integrated operational amplifier OP07 has the same function with F004. Consult Appendix 5 for the detail about the base pin of the two types of integrated operational amplifier.

Experiment

1. Join the circuit

(A) According to Figure 9 – 1 join integrated operational amplifier on the experimental circuit board.

(B) Join the DC power supply of $\pm 12V$ as shown in Figure 9 – 2.

(C) Notice

(a) The integrated operational amplifier may be damaged for inversely power poles, power accidental change or excessive high power.

(b) Power and input signal must have been cut off before any change of circuit, or the integrated operational amplifier may be damaged.

(c) The output end of integrated operational amplifier can not be connected with ground.

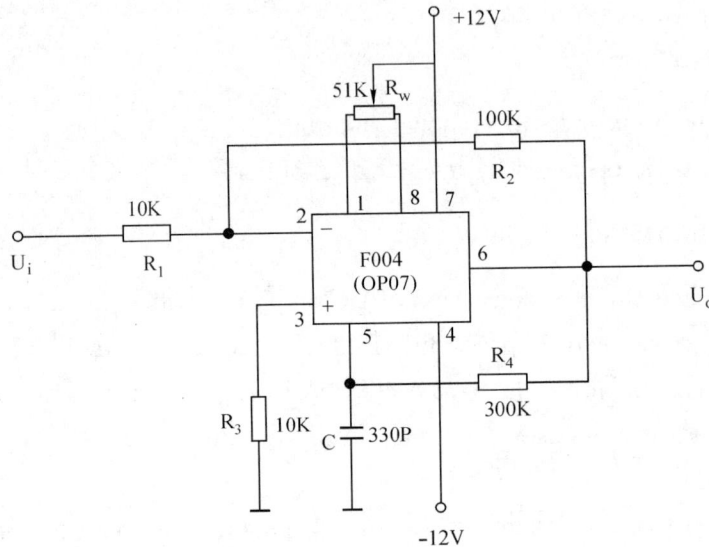

Figure 9 – 1

2. Zeroing

Connect the inverse input end of U_i with ground. Switch on power. Adjust R_W till the output voltage U_o becomes zero.

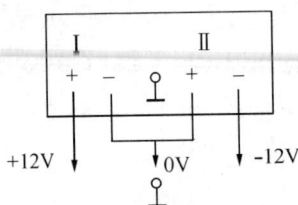

Figure 9 – 2

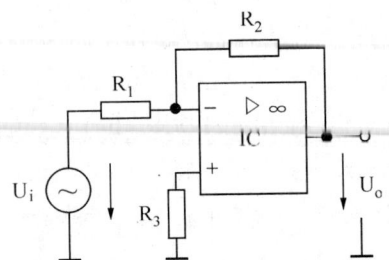

Figure 9 – 3

3. Inverse-phase proportional operation

(1) Figure 9 – 3 is the circuit of inverse-phase proportional operation constructed upon integrated operational amplifier. R_2 is a feedback resistance, R_1 is an external resistance. Thus, $U_o/U_i = -R_2/R_1$. So, for different values of R_2 or R_1 can get different ratios of U_o/U_i.

To keep the balance of two input ends in the circuit, the resistance R_3 should be $R_3 =$

$R_1 /\!/ R_2$.

(2) Calculation

For the circuit in Figure $9-2$, $R_1 = 10\mathrm{k}\Omega$, $R_2 = 100\mathrm{k}\Omega$, and R_3 equils $R_1 /\!/ R_2 = 9.1\mathrm{k}\Omega$ or approximately 10K. Then, $U_\mathrm{o} = \dfrac{R_2}{R_1} U_\mathrm{i} = -10 U_\mathrm{i}$.

(3) Measurement

Input a signal of $f = 10^3\mathrm{Hz}$ and $U_\mathrm{i} = 0.6\mathrm{V}$ on the inverse-phase end. Observe the input and output signals, and are they have the relation of $U_\mathrm{o} = -10 U_\mathrm{i}$? (Note: None of the elements in the circuit of Figure $9-3$ can be omitted.)

Signal	Voltage Value	$U - t$ Curve
U_i	0.6V	
U_o		

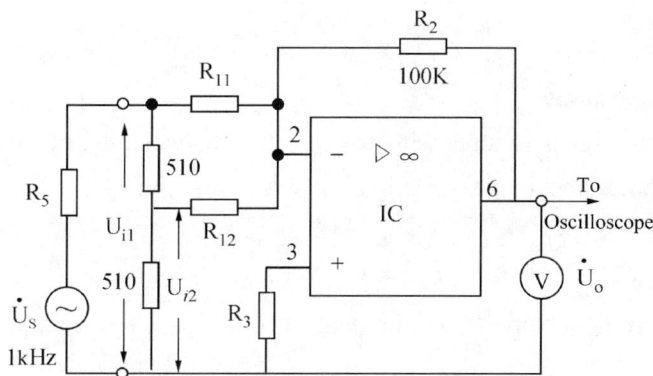

Fugure $9-4$

4. **Addition operation**

(1) Figure $9-4$ is the schematic diagram of addition operation. Two input signals U_i1 and U_i2 are combined together in certain proportion, and there are two resistances R_{11} and R_{12} connected at the input ends. Thus:

$$U_\mathrm{o} = -\left(\frac{R_2}{R_{11}} U_\mathrm{i1} + \frac{R_2}{R_{12}} U_\mathrm{i2} \right).$$

So, the circuit carried out addition operation.

(2) Let $R_2 = 100\mathrm{k}\Omega$, for the requirement of $U_\mathrm{o} = -(10 U_\mathrm{i1} + 10 U_\mathrm{i2})$, we can calculate that: $R_{11} = 10\mathrm{k}\Omega$, $R_{12} = 10\mathrm{k}\Omega$ and $R_3 = 5\mathrm{k}\Omega$.

(3) According to your calculation and the schematic diagram in Figure $9-4$, prepare the addition circuit on the experiment board.

(4) The input signal of $f = 10^3\mathrm{Hz}$ and $U_\mathrm{i} = 0.6\mathrm{V}$ will be divided into two parts by two $510\ \Omega$ resistances in series, and coupled to the two input ends of the circuit. Thus:

$U_{i1} = 0.6V$, $U_{i2} = 0.3V$. Measurement is the same as step 3. Is it true for the relation $U_o = (10U_{i1} + 10U_{i2})$?

Signal	Voltage Value	$U - t$ Curve
U_{i1}	0.6V	
U_{i2}	0.3V	
U_o		

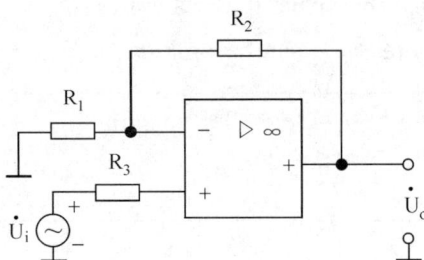

Figure 9 – 5

5. In-phase operation

(1) If the input signal is joint with the in-phase end of integrated operational amplifier, the circuit becomes the in-phase amplifier shown in Figure 9 – 5. The output voltage will be:

$$U_o = (1 + R_2/R_1) \ U_i$$

(2) Let $R_2 = 100k\Omega$, for the requirement of $U_o = 11U_i$, we can calculate that $R_1 = 10k\Omega$, $R_3 = 10k\Omega$.

(3) According to your calculation and the schematic diagram in Figure 9 – 5, prepare the in-phase operation circuit on the experiment board.

(4) For the input signal of $f = 10^3 Hz$ and $U_i = 0.6V$, check is it true of $U_o = 11U_i$?

Signal	Voltage Value	$U - t$ Curve
U_i	0.6V	
U_o		

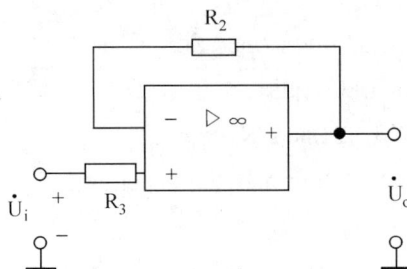

Figure 9 – 6

6. Voltage follower

(1) The circuit in Figure $9-6$ is called voltage follower. Join the circuit of voltage follower on the experiment board, let $R_2 = 10K$ and $R_3 = 10K$.

(2) For the input signal of $f = 10^3 Hz$ and $U_i = 0.6V$, check is it true of $U_o \approx 11 U_i$?

Signal	Voltage Value	$U - t$ Curve
U_i	0.6V	
U_o		

7. Integral operation

(1) The circuit in Figure $9 - 7$ is an integral operation circuit. In the integral circuit the feedback impedance is a capacitor and external impedance at the input end is a resistance.

(2) Join the integral operation circuit in Figure $9 - 7$ on the experiment board.

When the U_i is a rectangular wave, the U_o will be a saw-tooth wave, the saw-tooth wave is the integral of rectangular wave.

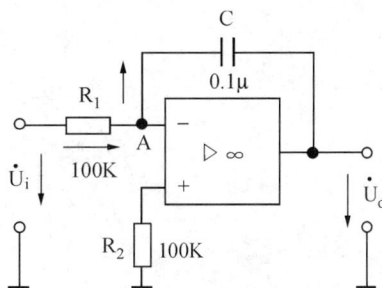

Figure $9 - 7$

(3) Input a square wave of peak-to-peak value $U_{iP-P} = 0.6V$. Observe the input and output waveform on the oscilloscope, and draw them down.

Signal	Waveform $U - t$
U_i	
U_o	

8. Differential operation

(1) The circuit in Figure $9 - 8$ is a differential operation circuit. In the differential circuit the feedback impedance is a resistance and external impedance at the input end is a capacitor.

Join the differential operation circuit in Figure $9 - 8$ on the experiment board.

When U_i is a rectangular wave, U_o will be two same narrow pulses, one is positive and the other is negative, which are the differential results of rectangular wave. In the same way, when U_i is a saw-tooth wave, the U_o will be rectangular wave.

(2) Input a square wave of peak-to-peak value

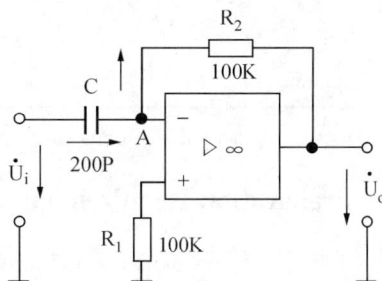

Figure $9 - 8$

$U_{\text{iP}-\text{P}} = 0.6\text{V}$. Observe the input and output waveform on the oscilloscope, and draw them down.

Signal	Waveform $U - t$
U_i	
U_o	

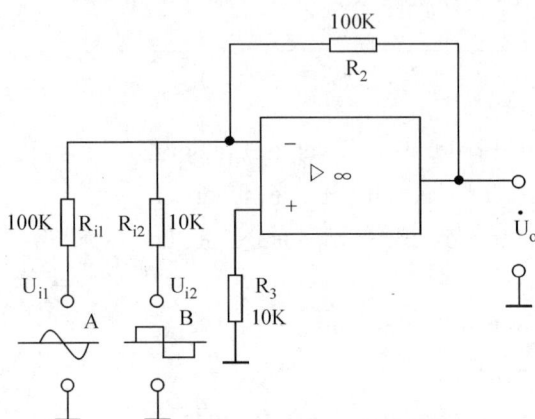

Figure 9 – 9

9. Scale adder

(1) In measuring apparatus, the integrated operational amplifier addition circuit is usually adopted as a scale adder, its circuit is shown in Figure 9 – 9. It is a circuit which adopts a sine wave as a standard to measure the period and width of a square wave.

(2) Join the scale adder circuit in Figure 9 – 9 on the experiment board.

(3) A standard sine wave of $f = 2000\text{Hz}$ and $U_{\text{i1P}-\text{P}} = 10\text{mV}$ from signal generator is input to the port A of the scale adder circuit. A square wave of $f = 100\text{Hz}$ and $U_{\text{i2P}-\text{P}} = 600\text{mV}$ is input to the port B of the circuit. Observe the waveform of U_{i1}, U_{i2} and U_o, and measure the period T of the square wave.

Signal	Voltage	$U - t$ Curve
U_{i1}	10 mV	
U_{i2}	600 mV	
U_o		

Measurement of the square wave: $T =$ _____ ms; displaying value $T =$ _____ ms.

Preparation Requirement

1. Know the working principle of integrated operational amplifier, and the circuit characteristics of proportional operator, follower, integral operator and differential operator.

2. Know the base pins of integrated operational amplifier F004 and OP07, and their

main technical parameters.

3. Understand your experiment contents and process.

Report Requirement

1. Record the data and waveforms of operators of inverse-phase, proportional, in-phase, follower and adder.

2. Write down the input and output waveforms of the circuit of integral, differential and scale adder.

3. Write down your understanding about this experiment.

Equipments

1. Two-trace oscilloscope.

2. Signal generator.

3. Experiment board for integrated operational amplifier circuit.

4. Digital multimeter.

实验十　集成运算放大器在波形发生方面的应用

【实验目的】

学会利用集成运算放大器构成正弦波发生器、矩形波发生器、三角波发生器和施密特触发器，熟悉其工作原理与近似计算。

【实验原理与方法】

在自动化设备和电子仪器中，经常需要进行性能的测试和信息的传递，这些都离不开一定的波形作为测试的依据。在模拟系统中，经常遇到正弦波发生器、矩形波发生器、三角波发生器和施密特触发器等。当前，从低频到中频的范围内，高质量的这些波形，都是利用集成运算放大器来产生的。

集成运算放大器 5G24（F007、μA471），其主要参数：开环差模增益 100 ~ 106db，最大输出电压 ±13V，差模输入电阻 2MΩ，共模抑制比 86dB。它的外引线名称：2 - 反相端，3 - 同相端，6 - 输出端，7 - 正电源，4 - 负电源，1、5 - 调零端。

集成运放的电源电压接反或电源电压突变、输入电压过大、输出端短路或过载时，都可能造成运放的损坏，所以在使用时应加保护措施。本实验因线路限制，未加保护电路。因此，在实验过程中，必须严格按照规定步骤进行，避免运放的损坏。

【实验内容和步骤】

1. 正弦波发生器

(1) 图 10 - 1 是 RC 桥式正弦波发生器的原理电路，它的振荡频率 $f = \dfrac{1}{2\pi RC}$。

(2) 根据原理图 10 - 1，连接电路，并计算发生器的振荡频率 f。

(3) 先不接入集成运放 5G24，测量各引线对地的电压，应是：

引线端	1	2	3	4	5	6	7	8
电压值	0	0	0	- 15V	0	0	+ 15V	0

(4) 断开电源，接入 5G24 后，再接通电源，测量各引线到地的电压。

外引线	1	2	3	4	5	6	7	8
电压值								

(5) 电路振荡 $f = f_0 = \dfrac{1}{2\pi RC}$，相移为 0，正反馈系数是最大 $F_V(+) = \dfrac{1}{3}$，在 f_0 附近，当 $F_V(-) < \dfrac{1}{3}$ 时，电路中的反馈是正反馈。观察是否 $F_V(-)$ 越接近 $\dfrac{1}{3}$，电路的

选频特性应越好，一般总是调整负反馈量，使之刚刚起振，再稍稍减小一点负反馈，输出即为不失真的正弦波。

（6）观察输出信号的波形，测量振荡周期 T 和振荡频率 f。

2．矩形波发生器

（1）图 10－2 是矩形发生器的原理电路，它的正反馈系数 $F = \dfrac{R_2}{R_1 + R_2}$，方波的周期 $T = 2R_f C\ln\left(1 + 2\dfrac{R_2}{R_1}\right)$，振荡频率 $f = \dfrac{1}{T}$。

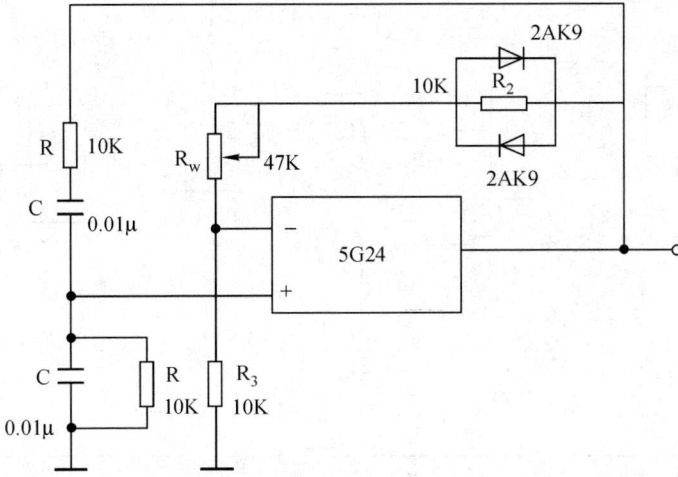

图 10－1

（2）根据图 10－2，连接电路，调节电位器 W_1，使 $R_1 = 18K$，调节 W_2，使 $R_f = 12.75K$，计算电路的正反馈系数 F，振荡周期 T 和频率 f。

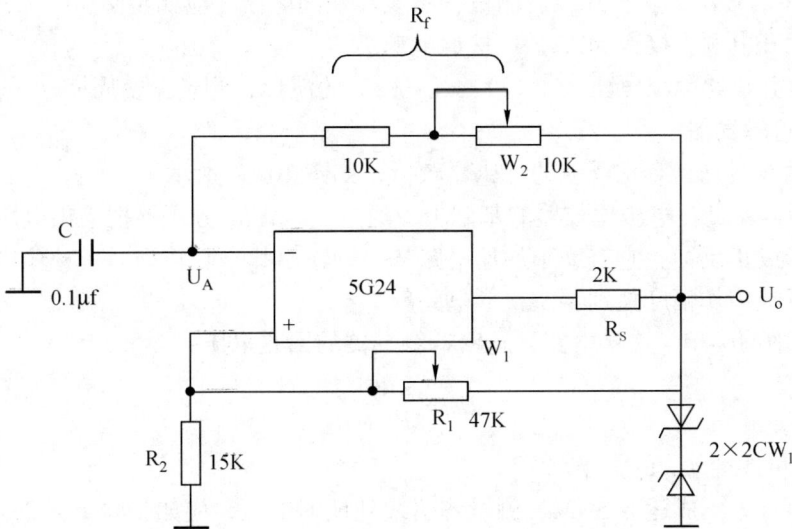

图 10－2

（3）先不接入集成运放 5G24，测量各引线对地的电压，应是步骤 1－（3）所示。

（4）先断开电源，接入 5G24，再接通电源。

（5）观察并描绘输出电压 $U_o(t)$ 和电容器 C 上的电压 $U_A(t)$ 的波形。

（6）测量输出电压 U_o 加到同相端的电压 U_+，正反馈系数 F，振荡周期 T 和振荡频率 f。

3．三角波发生器

（1）图 10－3 是三角波发生器的原理电路。

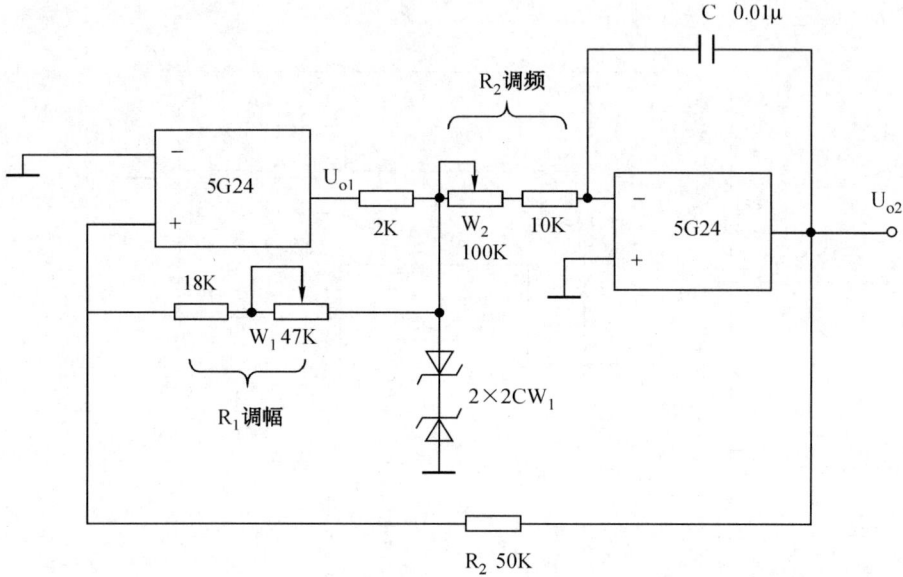

图 10－3

（2）根据图 10－3，连接电路，调节 W_1，使 $R_1 = 50K$，调节 W_2，使 $R_2 = 56.6K$。

（3）先不接入集成运放 5G24，测量各引线对地的电压，应如步骤 1－（3）所示。

（4）先断开电源，接入 5G24，再接通电源。

（5）观察并描绘输出电压 $U_{o1}(t)$，$U_{o2}(t)$ 的波形，测量振荡周期 T 和频率 f。

4．施密特触发器

施密特触发器是具有正反馈的电压比较器，如图 10－4 所示。

（1）图 10－4 是在电压比较器的基础上，通过 R_1 和 R_2 分压器把输出电压加到放大器的同相端而形成正反馈，它有助于加速电路从一种输出状态翻转到另一种输出状态，施密特触发器广泛地用作幅度甄别器及整形电路等。

（2）根据图 10－4，连接电路，施密特触发器的回差电压：

$$U_H = \frac{2R_2 U_o^+}{R_1 + R_2}$$

调节 W_1 可改变 U_H 的大小。

（3）先不接入集成运放 5G24，测量各引线到地的电压，应如步骤 1－（3）所示。

（4）先断开电源。接入 5G24 后，再接通电源。

（5）从信号发生器输入 $f = 100Hz$ 的正弦波信号，由双迹示波器观察 $U_i(t)$、$U_o(t)$ 的波形。

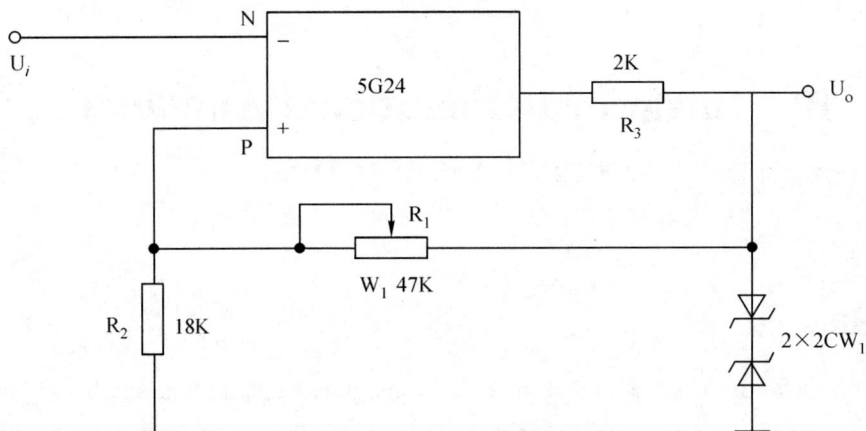

图 10 – 4

①测量 U_P，逐渐增大 U_i，记录刚出现矩形波时的 U_{im}，比较 U_N 和 U_P 值。

②描绘 $U_i(t)$、$U_o(t)$ 的波形。

（6）示波器水平显示开关置"X – Y"，将 U_i 输入示波器的 X 轴，U_o 输入 Y 轴，观察并描绘电路的传输特性，测量触发器的回差电压 U_H。

【实验预习要求】

1. 了解运算放大器的非线性特性，熟悉电压比较器、施密特触发器、正弦波发生器、矩形波发生器和三角波发生器的工作原理和电路特点。

2. 了解集成运放 5G24 外引线排列和主要技术参数。

3. 明确实验内容及其实施方法。

【实验报告要求】

1. 写出测试数据和波形。

2. 分析正弦波电路的选频特性、三角波电路的状态翻转和施密特电路的传输特性。

3. 将实验结果与计算值比较，分析产生误差的原因。

【实验仪器】

1. 双迹示波器 1 台。

2. 信号发生器 1 台。

3. 直流稳压电源 1 台。

4. 数字万用表 1 块。

5. 半导体线路实验仪 1 台。

10   Integrated Operational Amplifier:
Signal Generation

Objectives

Learn to make up of some signal generators integrated operational amplifier: sine wave, square wave, triangular wave and Schmidt trigger. Understand the working principle and the approximation of calculation.

Principle and Method

In automatic systems or equipments, there are some important processes such as characteristic testing and information transferring can not be carried out without the certain kinds of waves. Also in analog system, we often have to deal with sine wave generator, square wave generator, triangular wave generator and Schmidt trigger. Nowadays, in the range of low frequency to medium frequency, almost all of these high quality signals are generated by integrated operational amplifier.

The main parameters of integrated operational amplifier 5G24 (F007, μA471) are: open-loop differential-mode gain 100 ~ 106db, most output voltage $\pm$ 13V, differential-mode input resistance 2MΩ, common-mode rejection ratio 86dB. Its leg wires are: 2 — inversed phase end; 3 — in phase end; 6 — output end; 7 — positive end of power; 4 — negative end of power; 1 — and 5 — zeroing ends.

The integrated operational amplifier may be damaged for inversely power poles, power accidental change, excessive high power, overload and short circuit. So it should be protected in the circuit. In this experiment it is not protected for the limitation of lab source. Thus the experiment process must be dealt carefully to avoid damage.

Experiment

1. Sine wave generator

(1) Figure 10 − 1 shows the circuit if RC-bridge sine wave generator, its oscillatory frequency is: $f = \dfrac{1}{2\pi RC}$

(2) Join the circuit in Figure 10 − 1 and calculate the frequency f.

(3) Before connecting with integrated operational amplifier 5G24, check the voltages which may be led to the legs of it.

Leg wire	1	2	3	4	5	6	7	8
Voltage	0	0	0	$-15V$	0	0	$+15V$	0

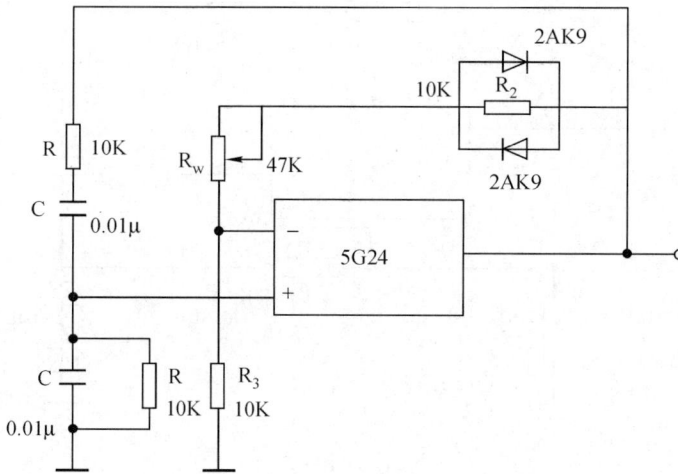

Figure 10 – 1

(4) Switch off power; plug in the 5G24. Then switch on power again, and test the voltages on the leg wires. Fill in the following table.

Leg wire	1	2	3	4	5	6	7	8
Voltage								

(5) For the oscillatory circuit, frequency is $f = f_0 = \dfrac{1}{2\pi RC}$, and phase drift is zero. The most positive feedback factor is $F_V\ (+)\ = \dfrac{1}{3}$, and near f_0 when $F_V\ (-)\ < \dfrac{1}{3}$ the feedback in the circuit is positive feedback. Observe whether the circuit acts as the nearer of $F_V\ (-)$ with $\dfrac{1}{3}$, the better of the frequency selection. Usually, we can adjust negatively feedback a little bit small just after the system oscillating to get an undistorted sine wave.

(6) Observe the output waveform and measure the oscillating period T and frequency f.

2. Square wave generator

(1) Figure 10 – 2 is a circuit schematic diagram of a square wave generator, its positive feedback factor is $F = \dfrac{R_2}{R_1 + R_2}$, the square wave period is $T = 2R_f C \ln\ (1 + 2\dfrac{R_2}{R_1})$, and oscillating frequency is $f = \dfrac{1}{T}$.

(2) Join the circuit according to Figure 10 – 2. Adjust variable resistance W_1 to $R_1 =$ 18K, and W_2 to $R_f = 12.75K$. Calculate the circuit positive feedback factor F, oscillating period T and frequency f.

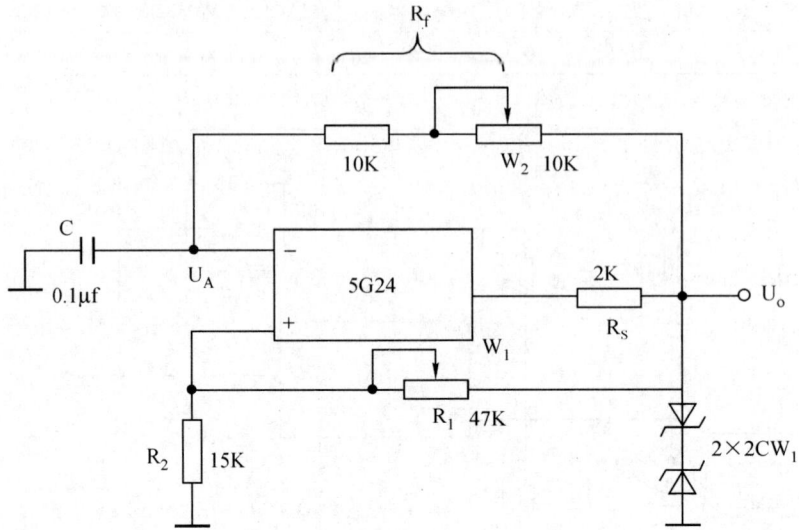

Figure 10 – 2

(3) Repeat the voltage checking process as it in last step.

(4) First, switch off power; then plug in the 5G24.

(5) Observe and describe the output voltage U_o (t) and the voltage U_A (t) on the capacitor C.

(6) Measure the output voltage U_o, voltage on the in-phase end U_+, oscillating period T and frequency f, calculate the positive feedback factor F.

3. Triangular wave generator

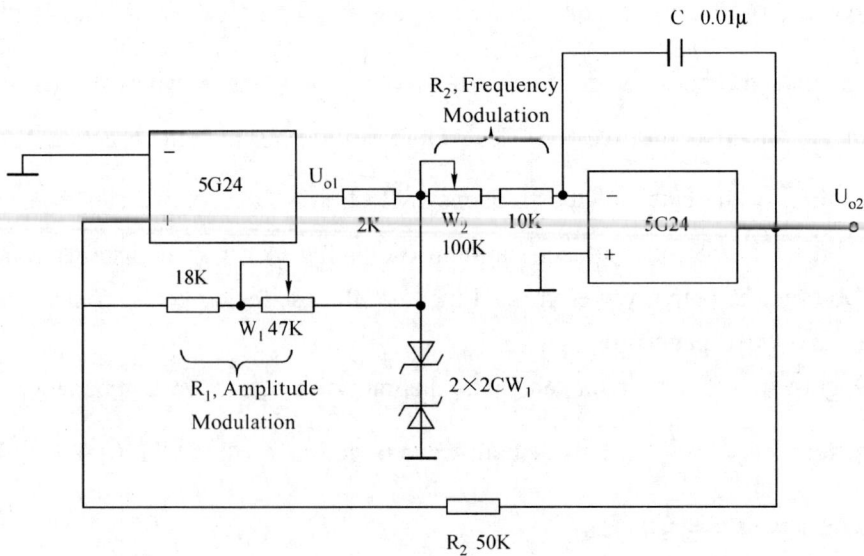

Figure 10 – 3

(1) Figure 10 – 3 is a circuit schematic diagram of a triangular wave generator.

(2) Join the circuit according to Figure $10-3$. Adjust variable resistance W_1 to $R_1 = 50K$, and W_2 to $R_2 = 56.6K$.

(3) Repeat the voltage checking process as it in last step.

(4) First, switch off power; then plug in the 5G24 and switch on power.

(5) Observe and describe the output voltages U_{o1} (t) and U_{o2} (t). Measure oscillating period T and frequency f.

4. Schmidt trigger

Schmidt trigger is a voltage comparator with positive feedback, as it shown in Figure $10-4$.

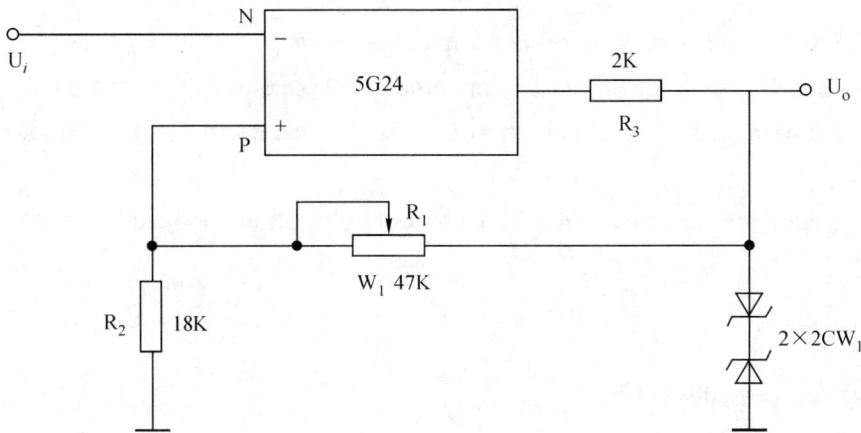

Figure $10-4$

(1) The Schmidt trigger circuit is constructed upon a voltage comparator, with a positive feedback of its output voltage connecting to the in-phase end through voltage divider R_1 and R_2. It may turn over from one output state to another output state rapidly, and is widely adopted as amplitude discriminate circuit or wave-shaping circuit.

(2) Join the circuit according to Figure $10-4$. The return difference voltage of Schmidt trigger is $U_H = \dfrac{2R_2 U_o^+}{R_1 + R_2}$, thus we can adjust variable resistance W_1 to change U_H.

(3) Repeat the voltage checking process as it in last step.

(4) First, switch off power; then plug in the 5G24 and switch on power.

(5) Input a sine wave of $f = 100Hz$, observe the waveform of U_i (t) and U_o (t).

① Increase U_i gradually. Record the voltage of U_{im} at the time of appearing of square wave, and compare the values of U_N and U_P.

② Describe the waveform of U_i (t) and U_o (t).

(6) Put the oscilloscope display key to the position "X − Y", input U_i to x-axis and U_o to y-axis. Observe and describe the transmission characteristics of the circuit, measure the return difference voltage U_H of Schmidt trigger.

Preparation Requirement

1. Know the nonlinear performance of the integrated operational amplifier, and the principle and characteristics of voltage comparator, Schmidt trigger, sine wave generator, square wave generator and triangular wave generator.

2. Know the base pins of integrated operational amplifier 5G24, and its technical parameters.

3. Understand your experiment contents and process.

Report Requirement

1. Record the data and waveforms in this experiment.

2. Understand the frequency selection process of sine wave generation circuit, state overturn of the triangular wave circuit and transmission characteristics of Schmidt trigger circuit.

3. Compare your experiment results with your calculation, and analyze the causation of errors.

Equipments

1. Two-trace oscilloscope.
2. Signal generator.
3. DC power supply.
4. Digital multimeter.
5. Experiment board for integrated operational amplifier circuit.

实验十一 直流稳压电路的调整和测试

【实验目的】

1. 掌握串联晶体管稳压电路的工作原理。
2. 学会测量稳压系数、纹波因数和电源内阻的方法。

【实验原理与方法】

典型的串联型晶体管稳压电路如图 11-1 所示。

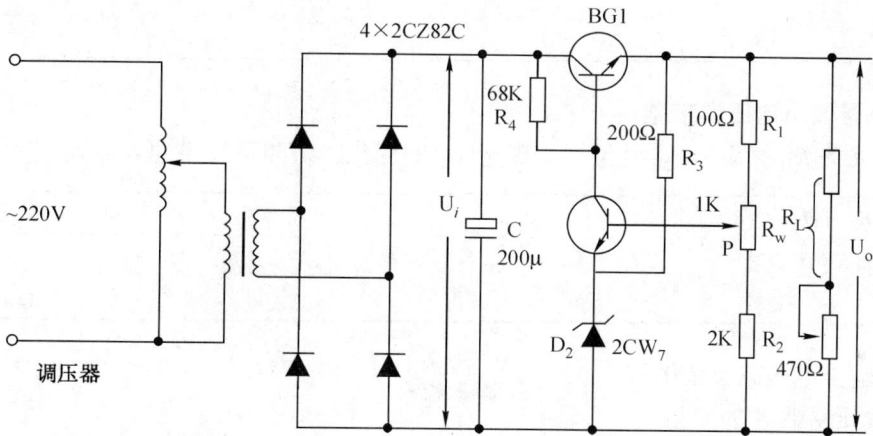

图 11-1

图中 BG_1 是调整管，BG_2 是放大管，电阻 R_1、R_2 和电位器 R_W 取出电压负反馈信号，硅稳压管 D_2 两端的稳定电压作为基准电压，负反馈信号电压与基准电压相比较，它们的差值就是 BG_2 的输入信号电压。R_3 是为硅稳压管提供电流通路并起限流作用的电阻，R_4 是 BG_2 的集电极电阻，同时也是 BG_1 的偏流电阻，C 是整流滤波电容。

电路的稳压原理可以简单地表示如下：设输入电压增加（或负载电流减小）而使输出电压增加时，其稳压过程：

$$U_o \uparrow \rightarrow U_{B2} \uparrow \rightarrow I_{C2} \uparrow \rightarrow U_{C2} \downarrow \rightarrow U_{B1} \downarrow \rightarrow U_{CE1} \uparrow \rightarrow U_o \downarrow$$

同理，当输入电压减小（或负载电流增加）使 U_o 减小时，通过类似过程，使调整管的 U_{CE1} 减小，也将使 U_o 基本保持不变。稳压电源稳压系数定义为：

$$S = \left. \frac{\Delta U_o / U_o}{\Delta U_i / U_i} \right|_{R_L = \text{Constant}}$$

稳压电源的内阻定义为：

$$R_o = \left. \frac{\Delta U_o}{\Delta I_o} \right|_{U_i = \text{constant}}$$

纹波因数 γ 为

$$\gamma = \frac{交流分量的总有效值}{直流分量}$$

【实验内容和步骤】

1．连接电路

按图 11 - 1 电路在实验板上连接成串联型晶体管稳压电路，并将调压器置于输出电压的最低位置。

2．测量稳压电路的工作状态

检查接线无误后接入 220V 市电，缓慢增高调压变压器的输出电压至 15V，然后检查各级工作状态，通常用万用表测量各管的 U_{BE} 应约为 0.7V，证明稳压电路工作状态正常以后，记录各点电压。

U_i (V)	U_o (V)	U_{CE1}	U_{CE2}

3．测量输出电压调节范围

断开负载 R_L，调节 R_W 电位器，观察 U_o 的变化，将测量结果记入下表。

1. R_W 位置	2. U_i (V)	3. U_{CE1} (V)	4. U_o (V)
5．P 点处于最低位置时	6.	7.	8.
9．P 点处于最高位置时	10.	11.	12.

所以输出电压调节范围 =

4．稳压系数 S 的测量

负载电阻调节为 150Ω，缓慢调节变压器的输出电压 U_i 由 15V 变化 ± 10%，测量 U_i 和 U_o 的变化，并计算 S 值。

U_i (V)	13.5	15.0	16.5
U_o (V)			

$$S = \frac{\Delta U_o / U_o}{\Delta U_i / U_i}\Bigg|_{R_L = \text{constant}(150\Omega)}$$

5．测量电源内阻 R_o

保持 U_i = 15V 不变，改变负载电阻 R_L，使输出电流 I_o 在 0～60mA 内任意变化，测量相应的输出电压的变化。

$$R_o = \frac{\Delta U_o\ (\text{mV})}{\Delta I_o\ (\text{mA})}\Bigg|_{U_i = 15V} = \qquad (\Omega)$$

6．测量纹波因数 γ

保持 U_i = 15V，用示波器观察 U_i 和 U_o 的波形，用示波器或数字万用表测量输入端和输出端的纹波电压，（即交流分量部分的有效值）计算纹波因数。

U_i（V）	输入端纹波 U_i（mV）	γ_i	U_o（V）	输出端纹波 $U_o\sim$（mV）	γ

【实验预习要求】

1. 学习串联型稳压电路的工作原理。
2. 了解稳压电源的调整和 S、R_o 与 γ 的测定方法。
3. 理解 U_o、U_i 纹波电压的概念。

【实验报告要求】

1. 画出实验原理线路图。
2. 列出所测数据，计算 S、R_o 和 γ。
3. 分析实验结果。

【实验仪器】

1. 双迹示波器 1 台。
2. 直流稳压电路实验板 1 块。
3. 数字万用表 1 块。
4. 调压变压器 1 台。

11　Debug of Circuit of DC Regulated Voltage

Objectives

1. Understand working principle of circuit of transistor-in-series regulated voltage.

2. Learn to measure the stability coefficient, ripple factor and the internal resistance of a power supply.

Principle and Method

Figure 11 − 1 shows a typical circuit of transistor-in-series regulated voltage.

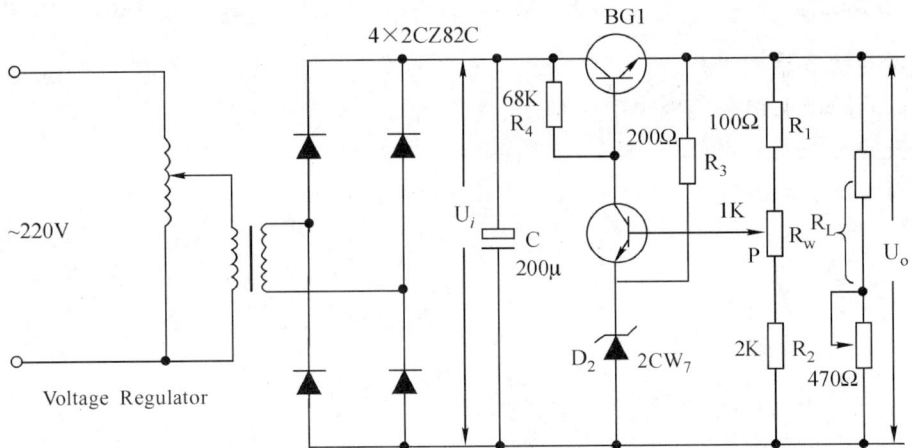

Figure 11 − 1

In the circuit, the BG_1 is a adjusting transistor, BG_2 is an amplifying transistor. A negative feedback voltage is taken by resistance R_1, R_2 and R_W, the voltage between the two ends of silicon Zener diode is choose as an standard voltage. The difference between feedback voltage and standard voltage is input into BG_2. Resistance R_3 acts as a current channel for the silicon Zener diode and current-limiting resistance. R_4 is the collector resistance for transistor BG_2, and also acts as a current bias resistance for BG_1. C is a rectifying and filtering capacitor.

The voltage regulating process can be shown as following: say the input voltage increasing (or the load current decreased), leads the output voltage U_o increased, and finally the output is regulated:

$$U_o \uparrow \rightarrow U_{B2} \uparrow \rightarrow I_{C2} \uparrow \rightarrow U_{C2} \downarrow \rightarrow U_{B1} \downarrow \rightarrow U_{CE1} \uparrow \rightarrow U_o \downarrow$$

In the same way, when the input voltage decreasing (or the load current increased), leads output voltage U_o decreased, and also the U_{CE1} of the adjusting transistor decreased, and finally the output U_o is kept unchanged. The stability coefficient of circuit is defined as:

$$S = \frac{\Delta U_o / U_o}{\Delta U_i / U_i}\bigg|_{R_L = \text{constant}}.$$

The internal resistance of the regulated power supply is defined as:

$$R_o = \frac{\Delta U_o}{\Delta I_o}\bigg|_{U_i = \text{constant}}.$$

The ripple factor γ is defined as:

$$\gamma = \frac{\text{total effective value of AC component}}{\text{DC component}}.$$

Experiment

1. Join the circuit

Join the circuit of transistor-in-series regulated voltage as it shown in Figure 11 – 1, and remember to set the regulating transformer at lowest voltage position.

2. Estimate the working performance of the circuit

After checking, your circuit can be switch on power of AC 220V. Gradually increase the output of the regulating transformer to about 15V, and then check the U_{BE} of all the transistors in circuit. If the U_{BE} are about 0.7V, the circuit works normally. Record your data.

U_i (V)	U_o (V)	U_{CE1}	U_{CE2}

3. Measure the output voltage variable range

Disconnect load resistance R_L, adjust R_W and observe the change of U_o. Record your measurement.

Position of R_W	U_i (V)	U_{CE1} (V)	U_o (V)
Lowest Position of P			
Highest Position of P			

So, the output voltage variable range is _____ V ~ _____ V.

4. Measure the stability coefficient S of circuit

Adjust the load resistance to 150Ω, and gradually adjust the output U_i of regulating transformer from 15V for a change of about $\pm 10\%$. Measure the changes of U_i and U_o, and calculated the value of S.

U_i (V)	13.5	15.0	16.5
U_o (V)			

$$S = \frac{\Delta U_o / U_o}{\Delta U_i / U_i}\bigg|_{R_L = \text{constant}(150\Omega)}.$$

5. Measure the internal resistance R_o of the regulated power supply

Keep $U_i = 15V$ unchanged, adjust the load resistance R_L to make the output current I_o changes in the range of $0 \sim 60mA$ optionally, and measure the corresponding changes of output voltage. Thus, internal resistance R_o equals:

$$R_o = \frac{\Delta U_o \ (\text{mV})}{\Delta I_o \ (\text{mA})}\bigg|_{U_i = 15V} = \qquad (\Omega)$$

6. Measure the circuit ripple factor γ

Keep $U_i = 15V$. Observe U_i and U_o on the oscilloscope. Measure the ripples voltage of input and output ends (effective value of AC component), and then calculate ripples factor γ.

U_i (V)	Ripple Voltage of Input End U_i (mV)	γ_i	U_o (V)	Ripple Voltage of Output End U_o (mV)	γ_o

Preparation Requirement

1. Learn the principle of circuit of transistor-in-series regulated voltage.

2. Understand the process of voltage self-adjustment and the measuring method for S, R_o and γ.

3. Understand the concept of the ripple voltage.

Report Requirement

1. Draw the schematic diagram.

2. List all the measurement data, calculate S, R_o and γ.

3. Discuss your experiment results.

Equipments

1. Two-trace oscilloscope.

2. Experiment board for DC regulated voltage circuit.

3. Digital multimeter.

4. Regulating transformer.

实验十二 逻辑门电路

【实验目的】

1．测试门电路的逻辑功能。
2．用集成与非门构成与门、或门、非门和或非门电路。

【实验原理与方法】

基本逻辑关系有三种，即与逻辑、或逻辑和非逻辑。

与逻辑表达式：

$$L = A \cdot B \cdot C$$

或逻辑表达式：

$$L = A + B + C$$

非逻辑表达式：

$$L = \overline{A}$$

由基本逻辑关系出发，可以导出逻辑代数的其他一些运算规律，组成各种复合门电路，如与非门，或非门，与或非门电路等。

在实际集成电路制造中，常以与非门作为逻辑部件的基本单元（它的价格也往往最低），再以它为基础构成与门、或门和其他复合门电路。本实验中，除与非门电路采用专门的集成电路外，与门、或门、或非门电路均采用集成与非门为基本单元构成。

TTL集成门电路由于具有工作速度快、输出幅度大、产品种类多、不容易损坏等特点，广泛使用于实验电路中。本实验采用74LS系列TTL集成门电路，它的工作电源电压为$5V \pm 0.5V$，逻辑高电平$1 \geqslant 2.4V$；低电平$0 \leqslant 0.4V$。

实验中，集成与非门电路采用74LS00（2输入端四与非门），74LS20（4输入端双与非门），74LS40（4输入端双与非门），与或非门电路采用74LS65（4-3-2-2与或非门）和74LS54（2-3-3-2与或非门），它们各自的管脚引线排列见附录六《集成逻辑门电路逻辑图、逻辑表达式与外引线排列》。

【实验内容和步骤】

1．测试集成与非门电路逻辑功能

（1）熟悉逻辑电路实验箱的使用，弄清其电源U_{CC}（为+5V）插孔、发光二极管显示插孔（该孔输入高电平灯亮），高、低电平信号开关插孔的位置，及测量时的接线方法。

（2）在面板上插上74LS20或74LS40（插座内部电路的地端与学习机地线、电路U_{CC}端与学习机的+5V电源已经连接）。将其4个输入端分别与4路电平信号开关插孔连接，其输出端接至发光二极管的插孔，检查无误后开启电源。

（3）当电路输入端 A、B、C、D 为下述情况，观察其输出端的逻辑状态，并用万用表测量输出端的电位，将结果填入表 12 – 1。

表 12 – 1

输入端	输出端 L	
$ABCD$	逻辑状态	电位（V）
1111		
0111		
0011		
0001		
0000		

验证上述结果是否符合与非门的逻辑表达式：$L = \overline{A \cdot B \cdot C \cdot D}$

2．用集成与非门组成其他基本逻辑电路，并测试其逻辑功能

（1）组成非门电路

将 74LS00 中的一个与非门的两个输入端相连接，则成为一个非门，如图 12 – 1 所示。

$$L = \overline{A}$$

当输入端 A 为下列情况时，观察输出端的逻辑状态，并测量其电位，将结果填入表 12 – 2 中。

注意事项：

①两个或多个 TTL 与非门组成逻辑电路时，其输出端不允许并联，否则将破坏正常的逻辑关系。

②TTL 与非门多余输入端可以悬空，也可以与已用的输入端并联使用，还可以通过一个 2kΩ 电阻接 + 5V 电源。

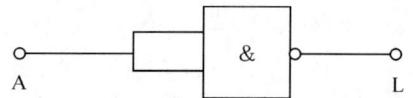

图 12 – 1

表 12 – 2

输入端	输出端 L	
A	逻辑状态	电位（V）
0		
1		

（2）组成与门电路

图 12 – 2

根据双重否定律，与门的逻辑代数式可作如下变换，$L = A \cdot B = \overline{\overline{A \cdot B}}$，因此可用两个与非门组成一个与门，如图 12 – 2 所示。

当输入端 A、B 下列情况时，将输出端的逻辑状态和电位值，填入表 12 – 3。

表 12 – 3

输入端		输出端 L	
A	B	逻辑状态	电位（V）
0	0		
0	1		
1	0		
1	1		

（3）组成或门电路

根据双重否定律和反演律，或门的逻辑代数式可作如下变换：

$$L = A + B = \overline{\overline{A + B}} = \overline{\overline{A} \cdot \overline{B}},$$

因此可用三个与非门组成一个或门，如图 12 – 3 所示。

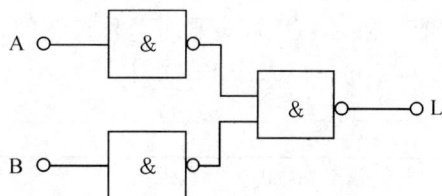

当输入端 A、B 为下列情况时，将输出端的逻辑状态和电位值，填入表 12 – 4 中。

图 12 – 3

表 12 – 4

输入端		输出端 L	
A	B	逻辑状态	电位（V）
0	0		
0	1		
1	0		
1	1		

（4）组成或非门

根据反演律和双重否定律，或非门的逻辑代数或可作如下变换：

$$L = \overline{A + B} = \overline{\overline{A} \cdot \overline{B}} = \overline{\overline{\overline{A} \cdot \overline{B}}},$$

因此，可用四个与非门构成一个或非门，如图 12 – 4 所示。

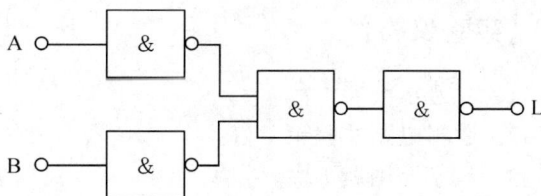

当输入 A、B 为下列情况时，将输出端的逻辑状态和电位值，填入表 12 – 5 中。

图 12 – 4

表 12 – 5

输入端		输出端 L	
A	B	逻辑状态	电位（V）
0	0		
0	1		
1	0		
1	1		

3．观察与非门的控制作用

（1）在面板上插入一块与非门电路（74LS00），将与非门的一个输入端接连续脉冲正方波的输出插孔。另一输入端接电平讯号的输出插孔，如图 12-5 所示。用双迹示波器同时观察与非门输入端和输出端的波形。

（2）与非门的控制端 A 为"关门"或"开门"状态时，观察并记录输入端 B 和输出端 L 的波形（当 A 获得高电平时，称控制门处于开门状态；当 A 接到低电平时，称控制门处于关门状态），填入表 12-6 中。

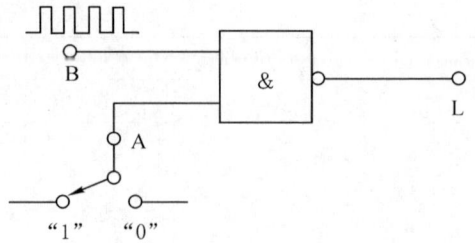

图 12-5

表 12-6

A 控制门状态	"0"（关门）	"1"（开门）
B 端输入波形		
L 端输出波形		

【实验预习要求】

1．复习逻辑门电路的工作原理和逻辑代数的基本定律。

2．与或非门电路中，不用的组和同一组中不用的输入端如何处理？

【实验报告要求】

1．整理实验数据，对数据及波形进行分析，解释实验中观察到的现象。

2．回答实验预习要求中的 2。

【实验仪器】

1．双迹示波器 1 台。

2．逻辑电路实验箱 1 台。

3．数字万用表 1 块。

12　Logic Gates Circuit

Objectives

1. Test the logic functions of logic gates circuit.
2. Construct AND, OR, NOT and NOR gates circuit by integrated NAND gates.

Principle and Method

There are three basic logic operations: AND, OR and NOT (the complement).

An expression containing AND operator: $L = A \cdot B \cdot C$

An expression containing OR operator: $L = A + B + C$

An expression containing NOT operator: $L = \bar{A}$

Basing on logic relations and operation rules, we can buildup some circuit of composite gates, such as: NAND, NOR, AOI (and-or-invert), etc.

Actually the NAND gate is a basic logic unit in the manufacture of integrated circuit, and its price is also lower. Usually NAND gates are basic block to build logic circuit of AND gate, OR gates and other composite gates. In this experiment we will construct our logic circuit all by NAND gates.

The characteristics of TTL integrated circuit is that rapid working speed, large output range, lots of products categories and resisting damage, so TTL integrated circuits are employed widely application. In this experiment, 74LS series TTL integrated circuits gates are adopt, its working source voltage is $5V \pm 0.5V$, logic 1 of higher voltage range $\geqslant 2.4V$, logic 0 of lower voltage range $\leqslant 0.4V$.

The NAND integrated circuit gates are supplied by 74LS00 (2 input ends 4 NAND gates), 74LS20 (4 input ends 2 NAND gates) and 74LS40 (4 input ends 2 NAND gates). The AOI integrated circuit gates are supplied by 74LS65 ($4 - 3 - 2 - 2$ AOI gates) and 74LS54 ($2 - 3 - 3 - 2$ AOI gates). Their base pins are listed in the Appendix 6.

Experiment

1. Test the logic functions of integrated circuit NAND gates

(1) Know the manipulation of the logic circuit box: jacks of source U_{CC} (+5V), jacks emitting LED (Light for higher voltage), jacks for higher and lower voltage switch, and wire connection.

(2) Plug in 74LS20 or 74LS40. Connect the 4 input ends with the 4 jacks of voltage

switches, the 2 output ends with jacks of emitting LED. Check your circuit and switch on power.

(3) Set different voltages for the input ends of A, B, C and D, observe logic states at their output ends and measure output voltages. Fill Table 12 – 1。

Table 12 – 1

Input	Output L	
ABCD	Logic state	Voltage (V)
1111		
0111		
0011		
0001		
0000		

Check your results with the logic NAND expression: $L = \overline{A \cdot B \cdot C \cdot D}$

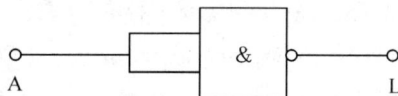

Figure 12 – 1

2. Logic circuits constructing by NAND gates and their logic functions

(1) Construct a circuit of NOT gate.

Connect two input ends of a NAND gate of 74LS00 to get a NOT gate, as shown in Figure 12 – 1.

$$L = \overline{A}$$

For different logic states of A, observe the logic output and measure its voltage. Fill Table 12 – 2.

Table 12 – 2

Input	Output L	
A	Logic state	Voltage (V)
1		
0		
1		

(2) Construct a circuit of AND gate.

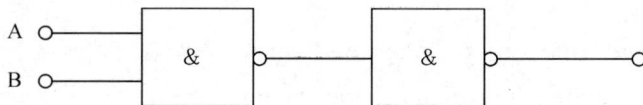

Figure 12 – 2

Because $L = A \cdot B = \overline{\overline{A \cdot B}}$, a AND gate can be built by two NAND gates, as shown in

Figure 12 – 2.

For different logic states of A and B, observe the logic output states and measure their voltages. Fill Table 12 – 3.

Table 12 – 3

Input		Output L	
A	B	Logic state	Voltage (V)
0	0		
0	1		
1	0		
1	1		

(3) Construct a circuit of OR gate.

According to logic algebra $L = A + B = \overline{\overline{A + B}}$ $= \overline{\overline{A} \cdot \overline{B}}$, a OR gate can be built by three NAND gates, as shown in Figure 12 – 3.

For different logic states of A and B, observe the logic output states and measure their voltages. Fill Table 12 – 4.

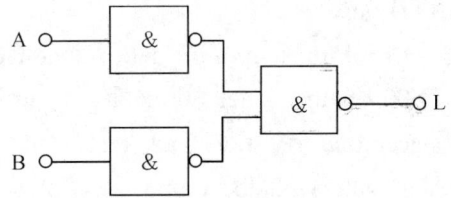

图 12 – 3

Table 12 – 4

Input		Output L	
A	B	Logic state	Voltage (V)
0	0		
0	1		
1	0		
1	1		

(4) Construct a circuit of NOR gate.

According to logic algebra $L = \overline{A + B} = \overline{A} \cdot \overline{B} = \overline{\overline{\overline{A} \cdot \overline{B}}}$, a NOR gate can be built by four NAND gates, as shown in Figure 12 – 4.

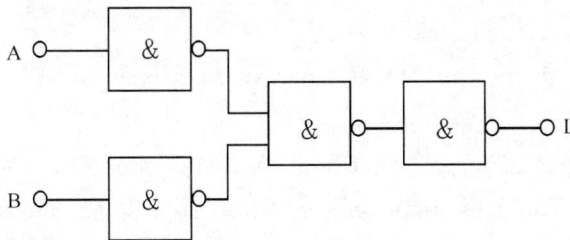

Figure 12 – 4

For different logic states of A and B, observe the logic output states and measure their voltages. Fill Table 12 − 5.

Table 12 − 5

Input		Output L	
A	B	Logic state	Voltage (V)
0	0		
0	1		
1	0		
1	1		

3. The controlling performance of a NAND gate

(1) Plug a NAND gate integrated circuit 74LS00 on the panel of the logic circuit box. Connect one of the input ends (B) of the NAND gate with the output jack of successive square pulses, another input end (A) with the jack of voltage level switch. See Figure 12 − 5.

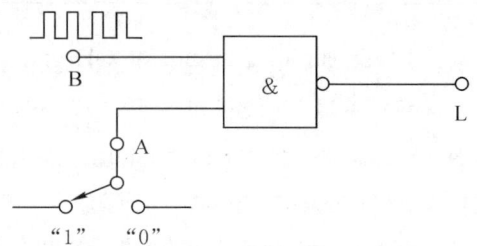

Figure 12 − 5

Observe the waveforms from input end and output end on the screen of oscilloscope.

(2) For states of "open" and "close" of the NAND gate control end A, observe the waveform from input end B and output end L. (When the NAND gate input end A is connected high voltage, the control gate is called open state, and when the NAND gate input end A is connected low voltage, the control gate is called close state.) Fill Table 12 − 6.

Table 12 − 6

State of Control gate A	"0" (Close)	"1" (Open)
Input Waveform at B		
Output Waveform at L		

Preparation Requirement

1. Review the logic circuit and logic algebra.

2. How to deal with the free NAND gates or free input ends?

3. Attention:

(1) The output ends of logic circuit built by two or more TTL NAND gates are not allowed to be in parallel, or the normal logic relationship will be damaged.

(2) The free input ends of TTL NAND gates can be hanged in air or connect with another input end in parallel, and also can be joined with +5V source through a 2kΩ resistance.

Report Requirement

1. Discuss the data and waveforms in this experiment, explain output results.
2. Answer question of last section.

Equipments

1. Two-trace oscilloscope.
2. Experimental box for logic circuit.
3. Digital multimeter.

实验十三　　双稳态触发器

【实验目的】

1．掌握基本 RS 触发器的组成、工作原理和性能。
2．熟悉集成 JK 触发器和 D 触发器的逻辑功能和测试方法。
3．掌握触发器间逻辑功能的转换。

【实验原理与方法】

　　触发器具有两个稳定状态，0 状态和 1 状态，在不同输入情况下，它可以被置于 0 状态，也可以被置于 1 状态。在输入信号消失以后，它能够保持状态不变。因此，用一个触发器，可以保存一位二进制信息。

　　集成触发器通过一些简单的连线或附加一些控制门，可以从一种功能的触发器转换到另一种功能的触发器。根据触发器的特性方程，可以方便地实现转换。

【实验内容与步骤】

1．基本 RS 触发器逻辑功能的测试

　　（1）从逻辑电路实验箱面板上插入一块（74LS00）双与非门，按图 13－1 所示电路接成基本 RS 触发器。

　　（2）$\overline{R}_D$、$\overline{S}_D$ 端分别接输出电平讯号插孔，利用电平的改变实现置"0"和置"1"；输出端 Q 和 $\overline{Q}$ 分别接发光二极管一端的插孔。

　　（3）当输入端 $\overline{R}_D$、$\overline{S}_D$ 为下列情况时，观察输出端 Q、$\overline{Q}$ 的逻辑状态，将实验结果填入表 13－1 中。

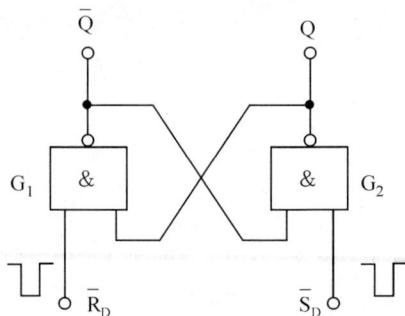

图 13－1

表 13－1

$\overline{R}_D$	$\overline{S}_D$	Q	$\overline{Q}$	触发器状态
0	1			
1	0			
1	1			
0	0			

2．集成 JK 触发器逻辑功能的测试

　　（1）选用 74LS112 双 JK 触发器插入面板，其外引线排列见附录六。

　　（2）置位和复位功能的测试，将 J、K、C 端开路，当 $\overline{R}_D$、$\overline{S}_D$ 分别为下列情况时，

观察输出端 Q 的逻辑状态，将结果填入表 13 – 2 中。

表 13 – 2

$\bar{R}_D$	$\bar{S}_D$	Q 端
0	1	
1	0	

（3）逻辑功能的测试。从 C 端输入单脉冲，将 J，K 分别接输出电平讯号插孔。先将触发器置 1 或置 0，在 $\bar{R}_D$、$\bar{S}_D$ 悬空的条件下，从 $C = 0$，$J = 0$，$K = 0$ 开始，按所列顺序，观察输出端的逻辑状态，注意是前沿触发还是后沿触发，并将实验结果填入表 13 – 3 中。

表 13 – 3

C	0	↑↓	0	↑↓	0	↑↓	0	↑↓
J	0	0	0	0	1	1	1	1
K	0	0	1	1	0	0	1	1
Q		1						
		0						

（4）将 JK 触发器的 J、K 端悬空，即 $J = K = 1$，然后从 C 端输入连续脉冲，用示波器观察 C 端的连续脉冲以及 Q 端的波形，并按图 13 – 2 予以记录。设触发器初态为 0。

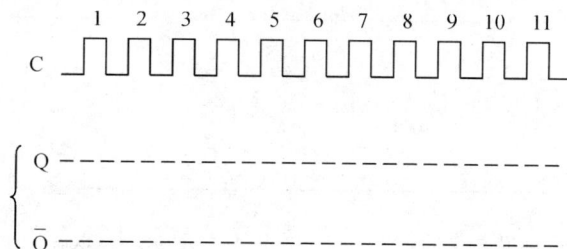

图 13 – 2

3. JK 触发器逻辑功能的转换

（1）JK 触发器转换为 T 触发器。

①将 JK 触发器中的 J、K 连接起来作为控制端时，就成 T 触发器，图 13 – 3 是 JK 触发器转换为 T 触发器的电路图。

②按图 13 – 3 连接电路，C 端加正单脉冲，观察 Q 端翻转次数和 C 端输入正单脉冲之间的关系，把结果填入表 13 – 4 中。

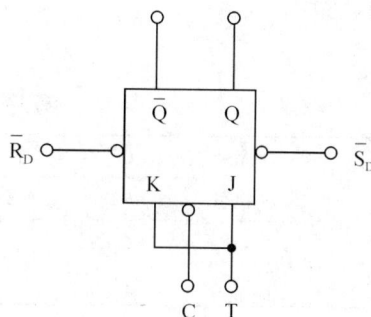

图 13 – 3

表 13-4

C	序号	0	1	2	3	4
	状态	0	↑↓	↑↓	↑↓	↑↓
Q 状态		0				
		1				

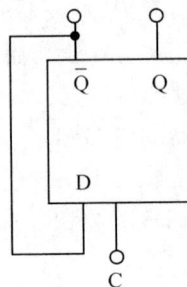

（2）JK 触发器转换为 D 触发器

① D 触发器的特性方程 $Q_{n+1} = D$，将此式与 JK 触发器的特性方程 $Q_{n+1} = \bar{J}Q_n + \bar{K}Q_n$ 比较可知，如取

$$J = D$$
$$K = \bar{D}$$

即可得到 D 触发器，转换电路如图 13-4 所示。

② 按图 13-4 连接电路，当 D 端为下列情况时，观察输出端 Q_n 到 Q_{n+1} 的逻辑状态，将结果填入表 13-5 中，检查输出状态与 D 端状态是否相同。

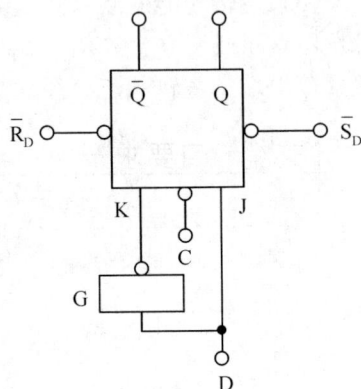

图 13-4

表 13-5

D	Q_n	Q_{n+1}	说明
0	0		
0	1		
1	0		
1	1		

（3）D 触发器转换为 T' 触发器

① 将 D 触发器 D 端和 $\bar{Q}$ 端连在一起，就成 T' 触发器，如图 13-5 所示。按图连接电路，C 端接正单脉冲输出端，观察 C 端输入正单脉冲后 Q 端的逻辑状态，将结果填入表 13-6 中。

表 13-6

C	序号	0	1	2	3	4
	状态	0	↑↓	↑↓	↑↓	↑↓
Q 状态		0				
		1				

图 13-5

②仍用上述电路，C 端输入连续正脉冲，用示波器观察 C、Q 或 $\overline{Q}$ 端的波形，并按图13-6予以记录（设触发器初态为0）。

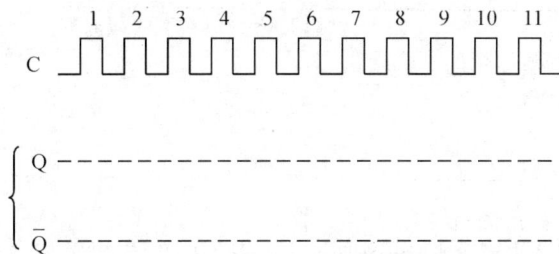

图 13-6

【实验预习要求】

1. 复习各种触发器的电路结构、工作原理和逻辑功能和触发方式。
2. 熟悉各种触发器逻辑功能的转换。

【实验报告要求】

1. 整理实验数据、波形，分析实验结果，总结各触发器的逻辑功能和特点。
2. 回答问题：
① $\overline{S}_D$ 和 $\overline{R}_D$ 两个输入端起什么作用？
②将 JK 触发器的 J、K 端悬空（称为 T' 触发器），试分析其逻辑功能。

【实验仪器】

1. 逻辑电路实验箱 1 台。
2. 双迹示波器 1 台。
3. 数字万用表 1 块。

13　Bistable Flip-flop

Objectives

1. Master the structure, working principle and performance of RS flip-flop.
2. Understand the logic functions and testing method of JK flip-flop and D flip-flop.
3. Master the logic function transfer of flip-flops.

Principle and Method

A flip-flop has two stable states. It can be set in state "0" or "1" under different input signals. It also can keep the output states after disappearing of the input signals. Thus, one flip-flop can save one digital binary signal or information.

With the helps of wire connection and additional controlling gates, the integrated circuit flip-flop can easily converse from one function to another function. The conversion can be realized theoretically under the help of characteristic equations of the flip-flops.

Experiment

1. Logic function test for basic RS flip-flop

(1) Plug a double-NAND-gate (74LS00) on the panel of logic circuit box; join the circuit as a RS flip-flop according to Figure 13-1.

(2) The driving ends $\bar{R}_D$ and $\bar{S}_D$ are connecting with jacks of voltage signal which can supply electric level of "0" and "1", the output ends of Q and $\bar{Q}$ are connecting with emitting LED.

(3) For different driving signals of $\bar{R}_D$ and $\bar{S}_D$ in the following table, observe the logic states of Q and $\bar{Q}$, and fill them into the table.

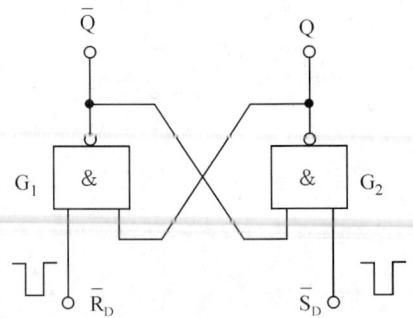

Figure 13-1

Table 13-1

$\bar{R}_D$	$\bar{S}_D$	Q	$\bar{Q}$	State of the Flip-flop
0	1			
1	0			
1	1			
0	0			

2. Logic function test for integrated

(1) Plug in a double JK flip-flop of 74LS112 on the panel of logic circuit box.

(2) Test the function of "Set" and "Reset". Let J, K and C be open circuit, for different driving signals of $\bar{R}_D$ and $\bar{S}_D$ in the following table, observe the logic state of Q, and fill them into Table 13 – 2.

Table 13 – 2

C	J	K	$\bar{R}_D$	$\bar{S}_D$	Q
×	×	×	0	1	
×	×	×	1	0	

(3) Test logic function. Connect J and K with jacks of voltage signal which can supply electric level of "0" and "1", input single pulse from C. Set the flip – flop "0" or "1". At first, set $\bar{R}_D$ and $\bar{S}_D$ free. Now begin from the state of $C = 0$, $J = 0$ and $K = 0$, following the order listing in Table 13 – 3, observe the logic state at the output end. Pay attention to triggering happens to the pulse leading edge or trailing edge. Fill Table 13 – 3.

Table 13 – 3

C	0	↑↓	0	↑↓	0	↑↓	0	↑↓
J	0	0	0	0	1	1	1	1
K	0	0	1	1	0	0	1	1
Q	1							
	0							

(4) Let the J and K ends be free, that means $J = K = 1$. Input successive pulses in the end C, observe the waveforms at the ends of C and Q on the oscilloscope. Record them in the Figure13 – 2. Suppose the initial state of the flip-flop is "0".

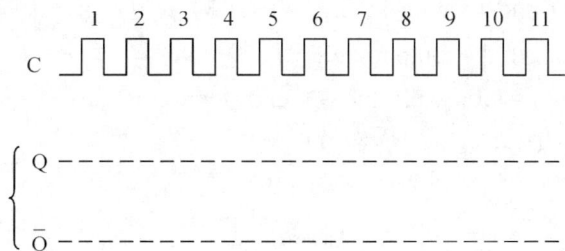

Figure 13 – 2

3. Logic function transfer of JK flip-flop

(1) JK flip-flop transfers into T flip-flop.

① Joining the two ends of J and K as a control end, the original JK flip-flop becomes

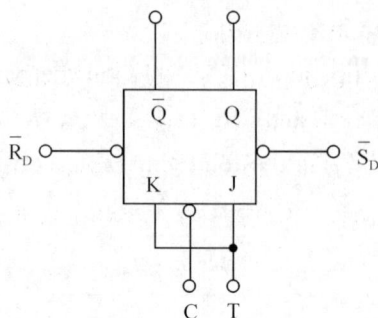

Figure 13 – 3

a T flip-flop, of which the circuit is shown in Figure 13 – 3.

② For the circuit in Figure 13 – 3, input single positive pulse. Observe the relation between the overturn times of the logic state of Q and the single positive pulse at the C end. Fill your results in Table 13 – 4.

Table 13 – 4

	Order	0	1	2	3	4
C	State	0	↑↓	↑↓	↑↓	↑↓
State of Q		0				
		1				

(2) *JK* flip-flop transfers into *D* flip-flop.

① The characteristic equation of D flip-flop is $Q_{n+1} = D$, and the characteristic equation of JK flip-flop is $Q_{n+1} = \bar{J}Q_n + \bar{K}Q_n$. Compare these two equations, let $J = D$ and $K = \bar{D}$, we can get a D flip-flop.

② Join the circuit in Figure 13 – 4. For the different logic states of the D end (listing in Table 13 – 5), observe the logic states on the output ends of Q_n and Q_{n+1}. Check the output state with the D end logic state, and fill the Table 13 – 5.

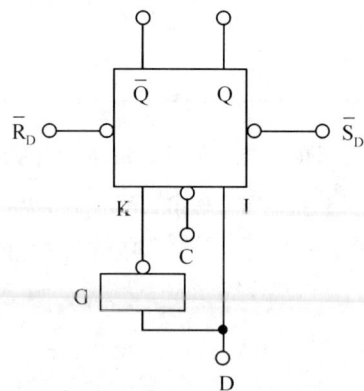

Figure 13 – 4

Table 13 – 5

D	Q_n	Q_{n+1}	Explaining
0	0		
0	1		
1	0		
1	1		

(3) D flip-flop transfers into T flip-flop.

① Join the D end with $\overline{Q}$ end of the D flip-flop, the D flip-flop transfers into a T flip-flop, showing in Figure $13 - 5$. Join the circuit, and input single positive pulse to end C. Observe the logic state of the end Q while the single positive pulse inputting at end C. Fill your results in Table $13 - 6$.

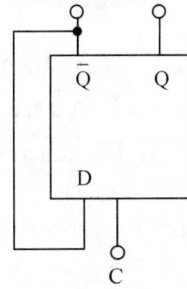

Figure $13 - 5$

Table 13 – 6

		Order	0	1	2	3	4
C		State	0	↑↓	↑↓	↑↓	↑↓
	State of Q		0				
			1				

② For the same circuit, input successive positive pulse at end C, observe the waveforms of C and Q or $\overline{Q}$, record your observation in Figure $13 - 6$. (Suppose the initial logic state is "0".)

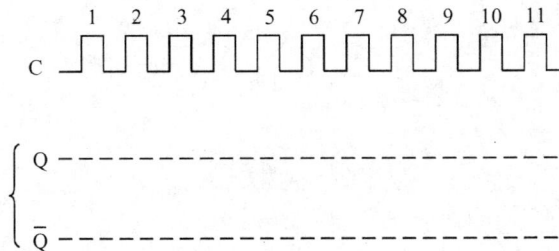

Figure $13 - 6$

Preparation Requirement

1. Review the knowledge about these kinds of flip-flops.
2. Understand logic function transfer of flip-flops.

Report Requirement

1. According to your experiment data and waveform. Analyze experiment results. Discuss the characteristics and logic functions of these kinds of flip-flop.

2. Answer questions:

(A) What are the working functions of the input ends $\overline{S}_D$ and $\overline{R}_D$?

(B) Hang the J and K ends of the JK flip-flop in air to built a T' flip-flop, try to discuss the logic function of it?

Equipments

1. Experimental box for logic circuit.
2. Two-trace oscilloscope.
3. Digital multimeter.

实验十四　逻辑代数的应用

【实验目的】

1．学会逻辑代数的应用。
2．掌握组合电路的实验分析方法。

【实验原理与方法】

利用逻辑代数的运算规律，可以用基本逻辑门（本实验把单片集成逻辑门电路视为基本逻辑门）构成具有一定逻辑功能的组合逻辑电路。反之，利用逻辑代数也可对所给组合逻辑电路进行逻辑分析。

本实验采用集成与非门、与或非门和集成异或门作为基本逻辑单元，分别按组合逻辑电路设计方法构成半加器和全加器。然后再用实验验证设计成的半加器与全加器的逻辑功能。

所谓半加器就是在进行二进制加法运算时，仅考虑本位的加数与被加数相加的加法逻辑电路；全加器则除考虑本位的加数、被加数外，还考虑由低位向本位的进位数。因此全加器的输出是三个数相加的结果。不论是半加器或全加器，其输出均包含两部分：留在本位的相加结果和数（用"S"表示，它取值为 0 或 1），以及向下一位的进位数（用 C 表示，取值也是 0 或 1）。

【实验内容和步骤】

1．用与非门组成半加器并测试其逻辑功能

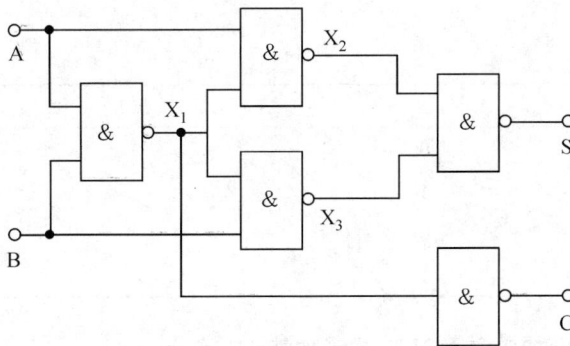

图 14 – 1

（1）用五个相同的与非门（2 块 74LS00）组成半加器，如图 14 – 1 所示。
（2）写出上图所示电路的逻辑代数表达式：

$X_1 =$

$X_2 =$

$X_3 =$

$C =$

$S =$

（3）按图 14 – 1 所示电路连线，检查无误后开启电源。

（4）当输入端 A、B 为下列情况时，观察 X_1、X_2、X_3、S 和 C 的逻辑状态，将结果填入表 14 – 1 中。

表 14 – 1

A	B	X_1	X_2	X_3	S	C
0	0					
0	1					
1	0					
1	1					

2．用异或门和与非门组成半加器，并测量其逻辑功能

（1）根据半加器的逻辑代数表达式可知，半加数 S 和 A，B 的关系是异或逻辑关系，而进位数 C 与 A，B 的关系是与逻辑关系。所以半加器也可用一个集成异或门和一个与门来实现，如图 14 – 2 所示。

（2）从面板插入异或门（74LS86）和与非门（74LS00），并按电路图接线（两个与非门可组成一个与门），经检查后开启电源。

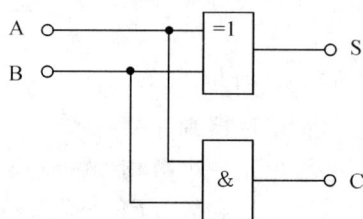

图 14 – 2

（3）当输入端 A、B 为下列情况时，观察输出端的逻辑状态，将实验结果填入表 14 – 2 中。

表 14 – 2

A	B	S	C
0	0		
0	1		
1	0		
1	1		

3．用与非门组成全加器并测试其逻辑功能

（1）由九个与非门组成的全加器，如图 14 – 3 所示。

（2）写出上图所示电路的逻辑代数表达式：

$Y =$ $Z =$

$X_1 =$ $X_2 =$

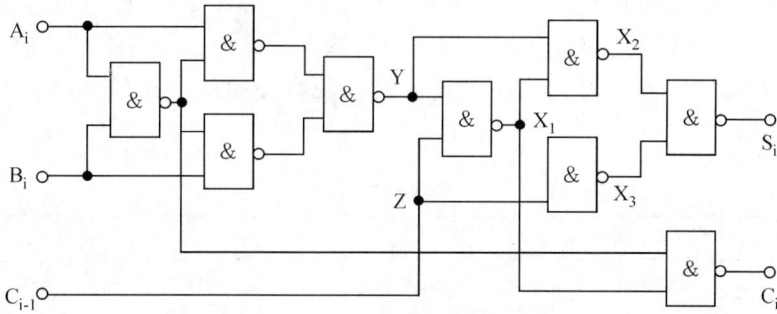

图 14 - 3

$X_3 =$ $C_i =$

$S_i =$

（3）按图 14 - 3 所示电路连线，经检查后开启电源。

（4）当输入端 A_i、B_i、C_{i-1} 为下列情况时，观察输出端的逻辑状态，并将实验结果填入表 14 - 3 中。

表 14 - 3

A_i	B_i	C_{i-1}	S_i	C_i
0	0	0		
0	1	0		
1	0	0		
1	1	0		
0	0	1		
0	1	1		
1	0	1		
1	1	1		

4. 用异或门和与非门组成全加器并测试其逻辑功能

（1）根据全加器的逻辑代数表达式可知，它可由两个异或门，一个与非门和一个非门组成，其逻辑如图 14 - 4 所示。

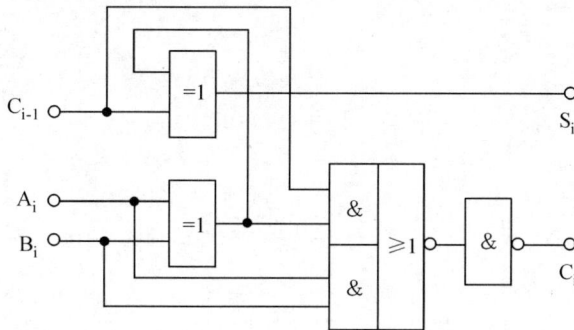

图 14 - 4

（2）在面板上插入四异或门（74LS86）、与或非门（74LS65）和与非门（74LS00），并按卜图电路接线，接线时把与或非门不用的与门输入端接地，经检查后开启电源。

（3）当输入端 A_i、B_i、C_{i-1} 为下列情况时，观察输出端的逻辑状态，并将实验结果填入表 14-4 中。

表 14-4

A_i	B_i	C_{i-1}	S_i	C_i
0	0	0		
0	1	0		
1	0	0		
1	1	0		
0	0	1		
0	1	1		
1	0	1		
1	1	1		

【实验预习要求】

1．复习组合逻辑电路的分析与设计基础。
2．了解半加器和全加器的工作原理。

【实验报告要求】

1．整理实验数据、图表，对实验结果进行分析探讨。
2．对由与非门组成的半加器和全加器，运用逻辑代数的基本定律，把逻辑式进行简化和变换，由最终的逻辑函数得到合理的逻辑电路图。
3．总结组合逻辑电路的分析方法。

【实验仪器】

1．逻辑电路实验箱 1 只。
2．数字万用表 1 块。

14   Application of Logic Algebra

Objectives

1. Learn to deal the logic function.
2. Master the analytical method for combination logic circuit.

Principle and Method

The utilization of logic algebra rules can realize some specific combination circuits which is built by basic logic gates. And on the other hand we can analyze combination circuit by making use of the logic rules. Usually, single chip of integrated gate circuit is called basic logic gate.

The basic logic gates in this experiment are integrated NAND gate, AOI gate and exclusive-OR gate. They could be the blocks to build half adder and full adder. And finally, we must approve the logic functions of the half adder and full adder in this experiment.

A half adder is a logic circuit for binary system addition operation, which only adds the addend and summand of the same bit. In contrary to a half adder, a full adder adds three numbers: the addend, summand and carry number from previous bit. Thus, the output of a full adder is the sum of three numbers. The outputs of half adder and full adder all have parts: the sum number of the self-bit (denoted as "S", may be "0" or "1".) and the carry number for next bit (denoted as "C" , may be "0" or "1".).

Experiment

1. Build a half adder and test its logic function

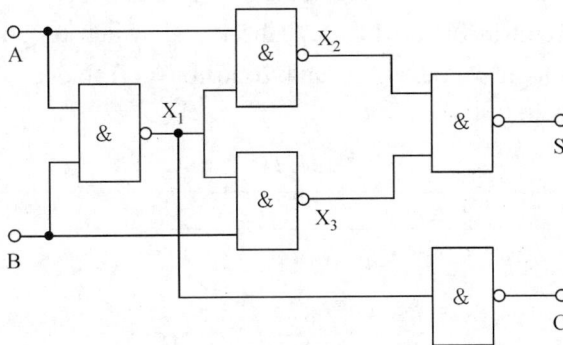

Figure 14 – 1

(1) Join 5 identical NAND gates (2 pieces 74LS00) into a half adder circuit, as shown in Figure14 − 1.

(2) Write the logic algebra expression of the circuit:

$X_1 =$

$X_2 =$

$X_3 =$

$C =$

$S =$

(3) Check your circuit, and then switch on power.

(4) For the input logic values of A and B listing in Table 14 − 1, observe the logic states of X_1, X_2, X_3, S and C. Fill in Table 14 − 1.

Table 14 − 1

A	B	X_1	X_2	X_3	S	C
0	0					
0	1					
1	0					
1	1					

2. Build a half adder by exclusive-OR gate and NAND gate, and test its logic function

(1) According to logic expression of a half adder, the half sum of S is a result of logic exclusive-OR operation of A and B, and carry C is a result of logic AND operation of A and B. Thus, a half adder can be realized by an integrated exclusive-OR gate and a AND gate, the circuit is shown in Figure 14 − 2.

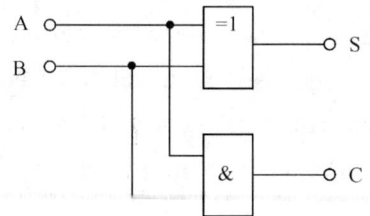

Figure 14 − 2

(2) Plug exclusive-OR gate (74LS86) and NAND gate (74LS00) on the panel of digit circuit box. Note that two NAND gates can build a AND gate. Join the circuit in Figure 14 − 2. Check and switch on power.

(3) For the input logic values of A and B listing in Table 14 − 2, observe the logic states of S and C. Fill in Table 14 − 2.

Table 14 − 2

A	B	S	C
0	0		
0	1		
1	0		
1	1		

3. Build a full adder by NAND gates and test it logic functions

(1) The circuit is in Figure 14 – 3; a full adder consists of 9 NAND gates.

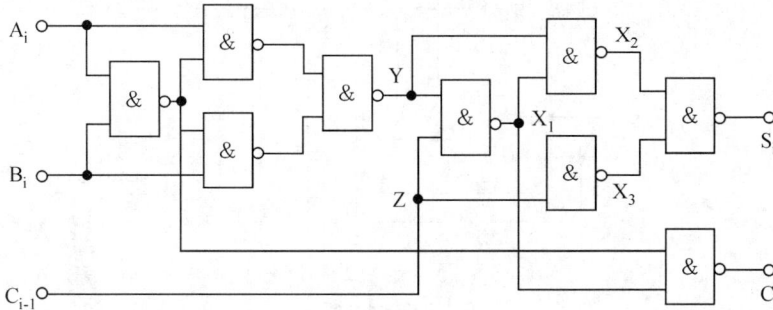

Figure 14 – 3

(2) Write the logic expressions of this circuit:

$Y =$ $Z =$

$X_1 =$ $X_2 =$

$X_3 =$ $C_i =$

$S_i =$

(3) Join the circuit. Check and then switch on power.

(4) For the input logic values of A, B and C_{i-1} listing in Table 14 – 3, observe the logic states of S_i and C_i. Fill in Table 14 – 3.

Table 14 – 3

A_i	B_i	C_{i-1}	S_i	C_i
0	0	0		
0	1	0		
1	0	0		
1	1	0		
0	0	1		
0	1	1		
1	0	1		
1	1	1		

4. Build a full adder by exclusive-OR gates and NAND gates, and test its logic functions

(1) The logic circuit of a full adder is in Figure 14 – 4; a full adder consists of two exclusive-OR gates, 1 NAND gates and 1 NOT gate.

(2) Plug a 4-exclusive-OR (74LS86) gate, AOI gate (74LS65) and NAND gate (74LS00) on the panel of digit circuit box. Join the circuit. Connect the spare input ends of AOI gate with ground. Check and switch on power.

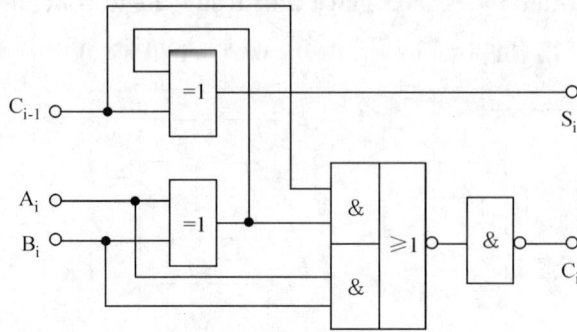

Figure 14 – 4

(3) For the input logic values of A, B and C_{i-1} listing in Table 14 – 4, observe the logic states of S_i and C_i. Fill in Table 14 – 4.

Table 14 – 4

A_i	B_i	C_{i-1}	S_i	C_i
0	0	0		
0	1	0		
1	0	0		
1	1	0		
0	0	1		
0	1	1		
1	0	1		
1	1	1		

Preparation Requirement

1. Review the knowledge about combination circuit, its analysis and designing.

2. The working processes of half adder and full adder.

Report Requirement

1. According to your experiment data and waveform, analyze experiment results. Discuss the characteristics and logic functions of the circuits.

2. According to basics logic rules, operate the logic expressions of half adder and full adder to a simple and rational form, and then realize it into actual circuit.

3. Summarize the lab methods for the combination circuit.

Equipments

1. Experimental box for logic circuit.

2. Digital multimeter.

实验十五　计　数　器

【实验目的】

1. 熟悉利用 JK 触发器组成计数器的方法。
2. 熟悉计数器的工作原理和逻辑功能的测试。

【实验原理与方法】

计数器的应用十分广泛，不仅可以计数，还可以用作数字系统中的定时电路和执行数字运算等。

计数器的种类繁多，本实验仅介绍使用最多的二进制计数器和 $8-4-2-1$ 编码的二－十进制计数器。

【实验内容和步骤】

1. 异步二进制加法计数器

（1）图 $15-1$ 是用 JK 触发器构成的异步二进制加法计数电路。在面板上插入 JK 触发器 74LS112，并按图 $15-1$ 所示连接线路，将各触发器的输出端 Q_3、Q_2、Q_1、Q_0 分别接发光二极管显示电路插孔。

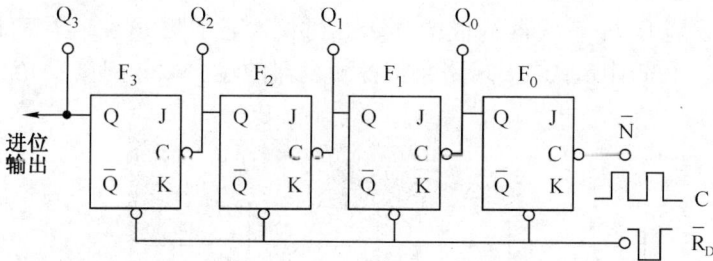

图 $15-1$　用主从 JK 触发器组成的异步二进制加法计数器

（2）计数开始前，首先清零。（方法是：将 $\bar{R}_D$ 端接"电平"插孔"0"，然后悬空）

（3）计数脉冲输入端接正单脉冲插孔，每输入一个计数脉冲后，观察各触发器输出端的逻辑状态，将实验结果填入表 $15-1$ 中。

表 $15-1$

输入脉冲	Q_3	Q_2	Q_1	Q_0
序号	2^3	2^2	2^1	2^0
0				
1				

输入脉冲 序号	Q_3 2^3	Q_2 2^2	Q_1 2^1	Q_0 2^0
2				
3				
4				
5				
6				
7				
8				
9				
10				
11				
12				
13				
14				
15				
16				

（4）将计数脉冲输入端接连续脉冲插孔，用示波器观察计数脉冲及各触发器 Q 端的输出波形，并对应地予以记录。

2．异步二进制减法计数器

（1）图 15－2 是用 JK 触发器构成的异步二进制减法计数电路。在面板上插入 JK 触发器（74LS112），并按图所示连接线路，将各触发器的输出端分别接发光二极管显示电路插孔。

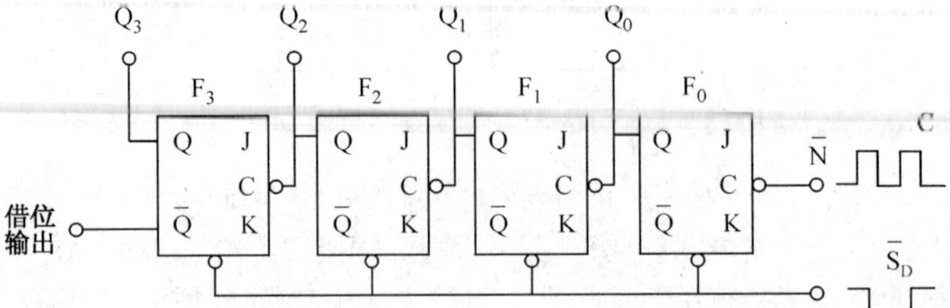

图 15－2

（2）计数开始前，首先使各触发器置 1（方法是，先将 $\overline{S}_D$ 端接电平插孔"0"，然后再悬空）。

（3）计数脉冲输入端接正单脉冲插孔，每输入一个计数脉冲后，观察各触发器输出端的逻辑状态，将实验结果填入表 15－2 中。

（4）将计数脉冲输入端连接到连续脉冲插孔，用示波器观察计数脉冲及各触发器 Q 端的输出波形，并对应予以记录。

表 15 – 2

输入脉冲	Q_3	Q_2	Q_1	Q_0
序号	2^3	2^2	2^1	2^0
0				
1				
2				
3				
4				
5				
6				
7				
8				
9				
10				
11				
12				
13				
14				
15				
16				

3. 同步二进制加法计数器

（1）图 15 – 3 是由 JK 触发器组成的同步二进制加法计数器，按图所示连接电路，（与门 G_1、G_2 可分别由两个与非门构成）将各触发器的输出端 Q_3、Q_2、Q_1、Q_0 分别接发光二极管显示电路插孔。

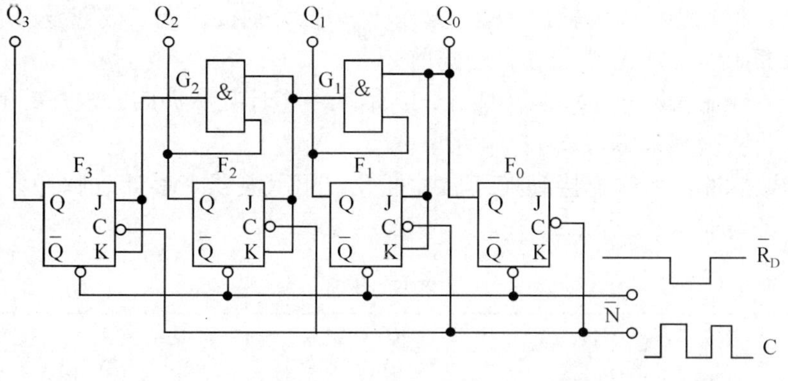

图 15 – 3　串行进位的四位同步二进制加法计数器

（2）计数开始前，首先清零。（方法同前）

（3）计数脉冲输入端接正单脉冲插孔，每输入一个计数脉冲后，观察各触发器输出端的逻辑状态，将实验结果填入表 15 – 3 中。

（4）将计数脉冲输入端接连续脉冲插孔，用示波器观察计数脉冲及各触发器 Q 端的输出波形，并对应地予以记录。

表 15 – 3

输入脉冲序号	Q_3 2^3	Q_2 2^2	Q_1 2^1	Q_0 2^0
0				
1				
2				
3				
4				
5				
6				
7				
8				
9				
10				
11				
12				
13				
14				
15				
16				

4．同步二进制减法计数器

（1）将图 15 – 3 中所有接到 Q 端连线改接至 $\overline{Q}$ 端，就构成了同步二进制减法计数器。改接成线路后，将各触发器的输出 Q_3、Q_2、Q_1、Q_0 分别接发光二极管显示电路插孔。

（2）计数开始前，首先置 1。（方法同前）

（3）计数脉冲输入端接正单脉冲插孔，每输入一个计数脉冲后，观察各触发器输出端的逻辑状态，将实验结果填入表 15 – 4 中。

（4）将计数脉冲输入端连接到连续脉冲插孔，用示波器观察计数脉冲各触发器 Q 端的输出波形，并对应地予以记录。

表 15 – 4

输入脉冲序号	Q_3 2^3	Q_2 2^2	Q_1 2^1	Q_0 2^0
0				
1				
2				

输入脉冲	Q_3	Q_2	Q_1	Q_0
序号	2^3	2^2	2^1	2^0
3				
4				
5				
6				
7				
8				
9				
10				
11				
12				
13				
14				
15				
16				

5．异步十进制加法计数器

（1）图 15 – 4 是由 JK 触发器构成的异步十进制加法计数器，按图示连接线路，将各触发器的输出端 Q_3、Q_2、Q_1、Q_0 分别接发光二极管的一端和数码显示 8421 端插孔，进位 C 接发光二极管显示电路插孔。

（2）计数开始前，首先清零。

（3）计数脉冲输入端接正单脉冲插孔，每输入一个计数脉冲后，观察各触发器输出端的逻辑状态和数码显示，将实验结果填入表 15 – 5 中。

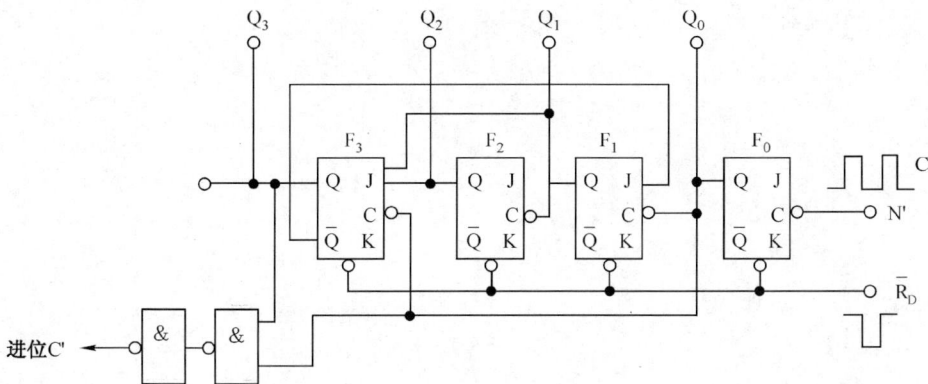

图 15 – 4   8421 码异步十进制加法计数器

表 15 – 5

态序 S	二进制状态				对应的十进制数	进位 C' 状态
	Q_3	Q_2	Q_1	Q_0		
0						
1						
2						
3						
4						
5						
6						
7						
8						
9						
10						

(4) 将计数脉冲输入端接连续脉冲插孔，用示波器观察计数脉冲及各触发器 Q 端的输出波形，并对应地予以记录。

【实验预习要求】

复习计数器的工作原理。

【实验报告要求】

整理实验数据和波形。

【实验仪器】

1. 逻辑电路实验箱 1 台。
2. 双迹示波器 1 台。
3. 数字万用表 1 块。

15　Counter

Objectives

1. Learn to build a counter by *JK* flip-flop.

2. Understand the working process of a counter and the test method for its logic functions.

Principle and Method

Counters are employed widely in different fields nowadays. It can not only count numbers, but also can work as a timing circuit in a number system, and manipulate digit operation.

There are a lot of kinds of counters. In this experiment, we will handle two kinds of counters: binary counter and binary coded decimal system $8 - 4 - 2 - 1$ coding counter.

Experiment

1. Asynchronous binary addition counter

(1) Figure $15 - 1$ shows the circuit of asynchronous binary addition counter, which is build by *JK* flip-flops. Plug *JK* flip-flops (74LS112) on the panel of digit circuit box. Join the circuit according to Figure $15 - 1$, and connect the output ends of Q_3, Q_2, Q_1 and Q_0 with jacks of the emitting LEDs.

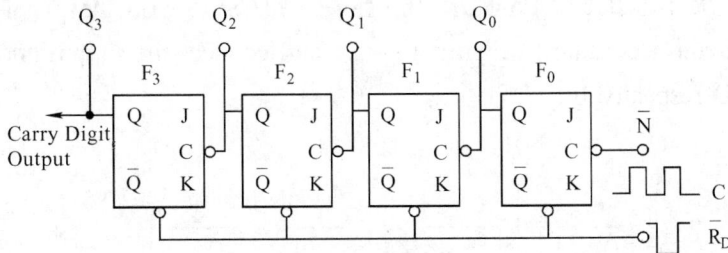

Figure $15 - 1$　Asynchronous binary addition counter
build by master-slave *JK* flip-flops.

(2) Before count number, do zero clearing first: touch the $\overline{R}_D$ end with voltage level of logic "0", and then let it hang in air.

(3) Connect the pulse input end of the counter with the jack of single positive pulse, and observe logic states of all the output ends of the counter after every pulse. Fill in Table $15 - 1$.

Table 15 – 1

Order of Input Pulses	Q_3 2^3	Q_2 2^2	Q_1 2^1	Q_0 2^0
0				
1				
2				
3				
4				
5				
6				
7				
8				
9				
10				
11				
12				
13				
14				
15				
16				

(4) Connect the pulse counting end with the jack of successive pulse, display the input pulses and output waveforms of Q end on the oscilloscope. Record them correspondingly.

2. Asynchronous binary subtraction counter

(1) Figure 15 – 2 shows the circuit of asynchronous binary subtraction counter, which is also build by JK flip-flops. Plug JK flip-flops (74LS112) on the panel of digit circuit box. Join the circuit according to Figure 15 – 2, and connect the output ends with jacks of the emitting LED respectively.

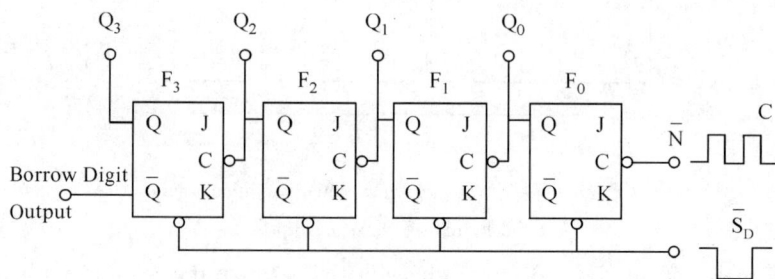

Figure 15 – 2

(2) Before count number, do one clearing first: touch the $\overline{S}_D$ end with voltage level of

logic "0", and then let it hang in air.

(3) Connect the pulse input end of the counter with the jack of single positive pulse, and observe logic states of all the output ends of the counter after every pulse. Fill in Table 15 − 2.

(4) Connect the pulse input end with the jack of successive pulse, display the input pulses and output waveforms of Q ends on the oscilloscope. Record them correspondingly.

Table 15 − 2

Order of Input Pulses	Q_3 2^3	Q_2 2^2	Q_1 2^1	Q_0 2^0
0				
1				
2				
3				
4				
5				
6				
7				
8				
9				
10				
11				
12				
13				
14				
15				
16				

3. Synchronous binary addition counter

(1) Figure 15 − 3 shows the circuit of synchronous binary addition counter, which is build by JK flip-flops. Plug JK flip-flops (74LS112) on the panel of digit circuit box. Join the circuit according to Figure 15 − 3, of which the two AND gates are made up by two NAND gates respectively, and connect the output ends of Q_3, Q_2, Q_1 and Q_0 with jacks of the emitting LEDs.

(2) Before count number, do zero clearing first, as the process previously.

(3) Connect the pulse input end of the counter with the jack of single positive pulse, and observe logic states of all the output ends of the counter after every pulse. Fill in Table 15 − 3.

(4) Connect the pulse input end with the jack of successive pulse, display the input pulses and output waveforms of Q ends on the oscilloscope. Record them correspondingly.

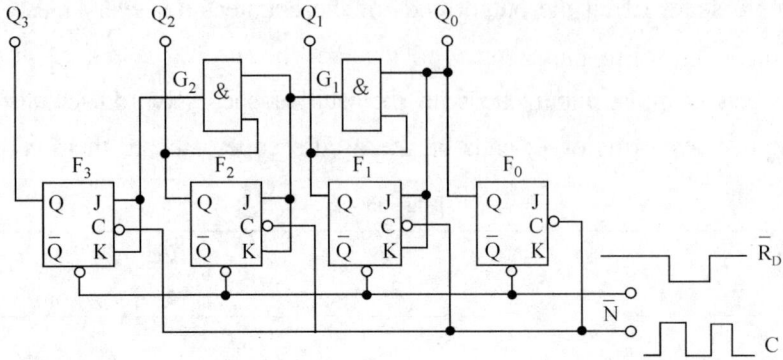

Figure 15 – 3   Four bits cascaded carry synchronous binary addition counter.

Table 15 – 3

Order of Input Pulses	Q_3 2^3	Q_2 2^2	Q_1 2^1	Q_0 2^0
0				
1				
2				
3				
4				
5				
6				
7				
8				
9				
10				
11				
12				
13				
14				
15				
16				

4. Synchronous binary subtraction counter

(1) As the circuit in Figure 15 – 3, all the wires connecting with Q ends must be altered connecting with $\bar{Q}$, and then we can get a synchronous binary subtraction counter. Connect the output ends of Q_3, Q_2, Q_1 and Q_0 with jacks of the emitting LEDs.

(2) Before count number, do one clearing first, as the process previously.

(3) Connect the pulse input end of the counter with the jack of single positive pulse, and observe logic states of all the output ends of the counter after every pulse. Fill in Table 15 – 4.

(4) Connect the pulse input end with the jack of successive pulse, display the input pulses and output waveforms of Q ends on the oscilloscope. Record them correspondingly.

Table 15 – 4

Order of Input Pulses	Q_3 2^3	Q_2 2^2	Q_1 2^1	Q_0 2^0
0				
1				
2				
3				
4				
5				
6				
7				
8				
9				
10				
11				
12				
13				
14				
15				
16				

5. Asynchronous decimal addition counter

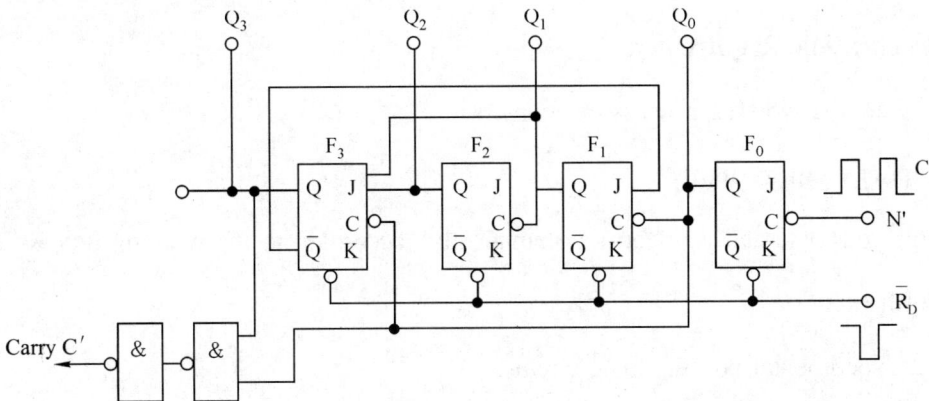

Figure 15 – 4   Asynchronous 8 – 4 – 2 – 1 coding decimal addition counter

(1) Figure 15 – 4 shows the circuit of synchronous decimal addition counter, which is

build by *JK* flip-flops. Plug *JK* flip-flops (74LS112) on the panel of digit circuit box. Join the circuit according to Figure 15 – 4. And connect all the output ends of Q_3, Q_2, Q_1 and Q_0 with jacks of the emitting LED, also with the jack of 8421 display indication tubes, the carry C connect with emitting LED.

(2) Before count number, do one clearing first, as the process previously.

(3) Connect the pulse input end of the counter with the jack of single positive pulse, and observe logic states of all the output ends and the displaying on LEDs after every pulse. Fill in Table 15 – 5.

Table 15 – 5

Order of States S	Binary States				Corresponding Decimal Number	State of Carry C'
	Q_3	Q_2	Q_1	Q_0		
0						
1						
2						
3						
4						
5						
6						
7						
8						
9						
10						

(4) Connect the pulse input end with the jack of successive pulse, display the input pulses and output waveforms of Q ends on the oscilloscope. Record them correspondingly.

Preparation Requirement

Review the working process of counters.

Report Requirement

Sort your data and waveform systematically, according to the working process.

Equipments

1. Experimental box for logic circuit.
2. Two-trace oscilloscope.
3. Digital multimeter.

附录一 常用低压熔丝规格及其额定电流

种类	直径(mm)	近似英规线号	额定电流(A)	种类	直径(mm)	近似英规线号	额定电流(A)	熔断电流(A)
	0.08	44	0.25		0.508	25	2	3.0
	0.15	38	0.5		0.559	24	2.3	3.5
	0.20	36	0.75		0.61	23	2.6	4.0
	0.22	35	0.8		0.71	22	3.3	5.0
	0.28	32	1		0.813	21	4.1	6.0
	0.29	31	0.05	铅锡合金丝（其中铜75%锡25%）	0.915	20	4.8	7.0
	0.36	38	1.25		1.22	18	7	10.0
	0.40	27	1.5		1.63	16	11	16.0
	0.46	26	1.85		1.83	15	13	19.0
	0.50	25	2		2.03	14	15	22.0
	0.54	24	2.25		2.34	13	18	27.0
青铝合金丝	0.58	23	2.5		2.65	12	22	32.0
	0.65	22	3		2.95	11	26	37.0
	0.94	20	5		2.26	10	30	44.0
	1.16	19	9		0.23	34	4.3	8.6
	1.26	18	8		0.25	33	4.96	9.8
	1.51	17	10		0.27	32	5.5	11.0
	1.66	16	11		0.32	30	6.8	13.5
	1.75	15	12.5		0.37	28	8.6	17.0
	1.98	14	15		0.46	26	11	22.0
	2.38	13	20	铜丝	0.56	24	15	33.0
	2.78	12	25		0.71	22	21	41.0
	3.14	10	30		0.74	21	22	43.0
	3.81	9	40		0.91	20	31	62.0
	4.12	8	45		1.02	19	37	78.0
	4.44	7	50		1.22	18	49	98.0
	4.91	6	60		1.42	17	65	125.0
	6.24	4	70		1.63	16	78	156.0
					1.83	15	96	191.0
					2.03	14	115	229.0

1 Staple Fuses and Their Rated Current

Sort	Diameter (mm)	Number in Britannic System	Rated Current (A)
Aluminum Alloy Wire	0.08	44	0.25
	0.15	38	0.5
	0.20	36	0.75
	0.22	35	0.8
	0.28	32	1
	0.29	31	0.05
	0.36	38	1.25
	0.40	27	1.5
	0.46	26	1.85
	0.50	25	2
	0.54	24	2.25
	0.58	23	2.5
	0.65	22	3
	0.94	20	5
	1.16	19	9
	1.26	18	8
	1.51	17	10
	1.66	16	11
	1.75	15	12.5
	1.98	14	15
	2.38	13	20
	2.78	12	25
	3.14	10	30
	3.81	9	40
	4.12	8	45
	4.44	7	50
	4.91	6	60
	6.24	4	70

Sort	Diameter (mm)	Number in Britannic System	Rated Current (A)	Breaking Current (A)
Lead-tin Alloy Wire (Lead 75%, tin 25%)	0.508	25	2	3.0
	0.559	24	2.3	3.5
	0.61	23	2.6	4.0
	0.71	22	3.3	5.0
	0.813	21	4.1	6.0
	0.915	20	4.8	7.0
	1.22	18	7	10.0
	1.63	16	11	16.0
	1.83	15	13	19.0
	2.03	14	15	22.0
	2.34	13	18	27.0
	2.65	12	22	32.0
	2.95	11	26	37.0
	2.26	10	30	44.0
Copper Wire	0.23	34	4.3	8.6
	0.25	33	4.96	9.8
	0.27	32	5.5	11.0
	0.32	30	6.8	13.5
	0.37	28	8.6	17.0
	0.46	26	11	22.0
	0.56	24	15	33.0
	0.71	22	21	41.0
	0.74	21	22	43.0
	0.91	20	31	62.0
	1.02	19	37	78.0
	1.22	18	49	98.0
	1.42	17	65	125.0
	1.63	16	78	156.0
	1.83	15	96	191.0
	2.03	14	115	229.0

附录二　功率表的使用

功率表是一种电动式仪表，可用于测量直流电路和交流电路的功率。它有两组线圈，一组是电流线圈，匝数较少，导线较粗，串接于被测电路中；另一组是电压线圈，匝数较多，导线较细，和附加电阻相联后并接在电路中。瓦特表的指针偏转是与电压、电流以及电压、电流这间的相位差角的余弦成正比的，即和被测电路的有功功率成正比，因此可用它测量电路的功率。

功率表的使用方法：

1．限量的选择

功率表通常都是多量限的，一般有两个电流量限。电流线圈一般用两个完全相同的绕组构成，使这两个绕组串联或并联，就可以转换电流量限。

功率表电压量限的改变和伏特表变换量限类似，是由电压线圈串联不同阻值的附加电阻来实现的。

在功率表中选用不同的电流和电压量限，就可以得到不同的功率量限。正确选择功率表的量限，实际上是要正确选择功率表中的电流量限和电压量限，不能仅从功率角度考虑。

例如，有一感性负载，功率约为 800W，额定电压 220V，功率因数 0.8，应怎样选择功率表的量限呢？我们可以这样考虑：因负载电压为 220V，故所选功率表的电压量限应为 250V 或 300V。负载电流 I 可按下式计算：

$$I = \frac{P}{U \cos\varphi} = \frac{800}{220 \times 0.8} = 4.54\text{A}$$

故可选用电流量限 5 安，这时功率量限为 $250 \times 5 = 1250\text{W}$，满足要求。

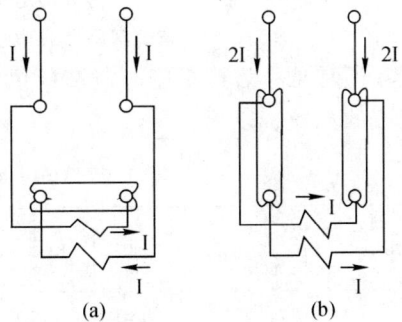

用连接片改变功率表的电流量限
(a) 两个电流绕组串联；
(b) 两个电流绕组并联，电流量限增加一倍

如果选用量限为 150V、10A 的瓦特表，功率量限为 1500W 满足要求，但负载电压超过了量限，电压线圈有可能损坏，所以不能这样使用。

2．功率表的正确接线

由于功率表的读数与电压、电流之间的相位差角有关，因此电流线圈、电压线圈的接线必须按照规定的方式接线，为了使接线不致发生错误，通常在电流线圈的一端（电流端）标有 ＊ 号的电流端接至电源相线，而另一端接至负载端。这时电流线圈是串联接入。标有"＊"号的电压端，接至 ＊ 号电流端钮的一端，而另一端则跨接到负载的另一端，这时电压线圈是并联接入电路的。如果不按这个规则接线，指针就会反转。

如果功率表的接线是正确的，但发现指针反转，则表明此时负载端实际上含有电源，反过来向外输出功率。若要想测出这个功率，可借助仪表上的电压线圈的"换向开关"，转动换向开关，就可以很方便地使指针朝正向偏转。

3．功率表的正确读数

一般功率表只标注分格数，而不标注功率数，这是由于功率表一般是多量限的，在选用不同的量限时，每一分格都代表不同的功率数称为功率表的分格常数。在测量时读得功率表的偏转格数，乘上功率表的相应分格常数，就等于被测功率的数值。

即：
$$P = C \cdot \alpha \text{（瓦）}$$

式中　P——被测功率的瓦数；

　　　C——功率表的分格常数（瓦/格）；

　　　α——指针偏转的格数。

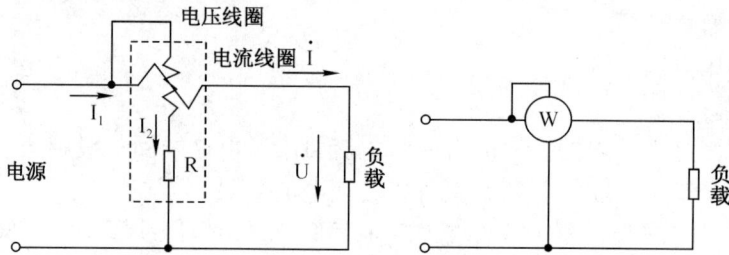

功率表接入被测电路，电流线圈
与负载串联，电压线圈与负载并联

功率表的符号

通常功率表内都附有制造厂供给的表格，注明在不同电流、不同电压量限下每一格所代表的瓦数，以供查用。

D34－W 刻度每格所代表之瓦特（C）

型号	电压 电流	25	50	100	50	100	200	75	150	300	150	300	600
	0.25 0.5	0.01 0.02	0.01 0.04	0.025 0.05	0.025 0.05	0.05 0.1	0.1 0.2	0.025 0.05	0.05 0.1	0.1 0.2	0.025 0.05	0.1 0.2	0.2 0.4
	0.5 1	0.025 0.05	0.05 0.1	0.05 0.1	0.05 0.1	0.1 0.2	0.2 0.4	0.05 0.1	0.1 0.2	0.2 0.4	0.1 0.2	0.2 0.4	0.4 0.8
D34－2	1 2	0.05 0.1	0.1 0.2	0.1 0.4	0.1 0.2	0.1 0.1	0.4 0.8	0.1 0.2	0.2 0.4	0.4 0.8	0.25 0.5	0.5 1	1 2
	2.5 5	0.1 0.2	0.2 0.4	0.4 0.8	0.25 0.5	0.5 1	1 2	0.25 0.25	0.5 1	1 2	0.5 1	1 2	2 4
	5 10	0.25 0.5	0.5 1	1 2	0.5 1	1 2	2 4	0.5 1	1 2	2 4	1 2	2 4	4 8

D26 – W 表　刻度每格所代表之瓦特（C）

额定电流（A）	额 定 电 压（V）						
	75	150	300	600	125	250	500
0.5	0.25	0.5	1	2	0.5	1	2
1	0.5	1	2	4	1	2	4
2	1	2	4	8	2	4	8
2.5	1.25	2.5	5	10	2.5	5	10
5	2.5	5	10	20	5	10	20
10	5	10	20	40	10	20	40
20	10	20	40	80	20	40	80

2　Power Meter

Power meter is a kind of electro motional meter which can be used to measure the power consumed in the DC or AC circuits. There are two coils in it: one is current coil which has less turns of thick wire and is connected in series with the measured circuit; the other is voltage coil which has more turns and is connected in parallel with the measured circuit. The deflection of the power meter pointer is direct proportional with voltage, current and the cosine of the phase angle between voltage and current, thus a power meter can measure active power.

Operation of a power meter:

1. Measure range

A power meter usually has multi measure ranges. There are two current coils which can be connected in series or in parallel to form two different current measure ranges. The converting of voltage measure range is realized by connecting in series with different resistances.

To select proper power measure range one should select different current and voltage measure range.

For example: an inductance load of power about 800W and rated voltage 220V, the power factor of circuit is 0.8. We can choose the voltage range as 250V or 300V, but the current I should be calculate according to following equation:

$$I = \frac{P}{U\cos\varphi} = \frac{800}{220 \times 0.8} = 4.54A$$

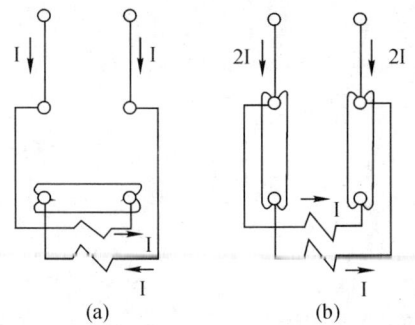

Convert current measure ranges by connecting flakes.
(a) The two coils in series　(b) The two coils in parallel, doubles the current measure range

Thus the current range should be 5A, and the power measure range will be $250 \times 5 = 1250W$, which is suitable for our measurement.

Again, if we choose ranges as 150V and 10A, which has a power range of 1500W smaller than the load power, but the load voltage exceed the range and can cause damage to voltage coil.

2. The connection of a power meter

Because the reading values are determined by the phase difference between current and voltage, the current and voltage coils should be connected in correct ways which are indicated by marks of " * " on the panel. The end of current coil which is marked by * should be connected with wire under voltage and the other end with load, and now the current coil is in series with the load. The end of voltage coil marked by * should be connected the same end of current coil and the other end connected with another end of load, and now the voltage coil is parallel with the load. If the meter is connected in a wrong way its pointer will deflect inversely.

If the meter is connected correctly but the pointer deflects inversely, there maybe power supply inside the load. To measure this power, one just turns the commutator switch on the panel and the pointer will deflect right.

3. Get correct readings of a power meter

Usually a power is marked with divisions, not power values, because it is a multi ranges meter. For different range, a division represents a different value of power, which is called division constant. To get a correct reading, one should read the deflecting divisions and time it with the division constant:

$$P = C \cdot \alpha \ (W)$$

Where P is the power value, C is division constant and α is the deflecting divisions.

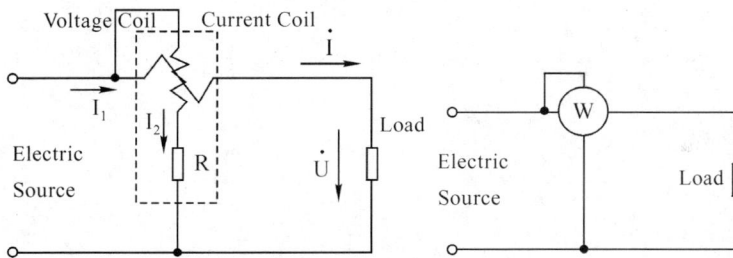

The connection of a power meter: the current coil is in series with load and the voltage coil is parallel with load.

The sign of a power meter in a circuit.

Usually there is a data table attaching with power meter which is provide by manufactory. Consult the table one can get the division constant under different measuring range.

Division Constant (C) of Power Meter Type D34 – W

Current	Voltage (V)											
(A)	25	50	100	50	100	200	75	150	300	150	300	600
0.25	0.01	0.01	0.025	0.025	0.05	0.1	0.025	0.05	0.1	0.025	0.1	0.2
0.5	0.02	0.04	0.05	0.05	0.1	0.2	0.05	0.1	0.2	0.05	0.2	0.4
0.5	0.025	0.05	0.05	0.05	0.1	0.2	0.05	0.1	0.2	0.1	0.2	0.4
1	0.05	0.1	0.1	0.1	0.2	0.4	0.1	0.2	0.4	0.2	0.4	0.8
1	0.05	0.1	0.1	0.1	0.1	0.4	0.1	0.2	0.4	0.25	0.5	1
2	0.1	0.2	0.4	0.2	0.1	0.8	0.2	0.4	0.8	0.5	1	2

Current (A)	Voltage (V)											
	25	50	100	50	100	200	75	150	300	150	300	600
2.5	0.1	0.2	0.4	0.25	0.5	1	0.25	0.5	1	0.5	1	2
5	0.2	0.4	0.8	0.5	1	2	0.25	1	2	1	2	4
5	0.25	0.5	1	0.5	1	2	0.5	1	2	1	2	4
10	0.5	1	2	1	2	4	1	2	4	2	4	8

Division Constant (C) of Power Meter Type D26 – W

Current (A)	Voltage (V)						
	75	150	300	600	125	250	500
0.5	0.25	0.5	1	2	0.5	1	2
1	0.5	1	2	4	1	2	4
2	1	2	4	8	2	4	8
2.5	1.25	2.5	5	10	2.5	5	10
5	2.5	5	10	20	5	10	20
10	5	10	20	40	10	20	40
20	10	20	40	80	20	40	80

附录三 兆欧表的使用

兆欧表是用来测量绝缘电阻的仪表。绝缘材料的绝缘性能会因受潮、发热、老化等原因而下降，当其达不到规定的要求时，设备就不能正常使用。

兆欧表主要由一台小容量，输出高电压的手摇发电机和一只磁电管比率表及测量线路组成，所以兆欧表又被称为摇表。兆欧表的外形如附图 3−1 所示。

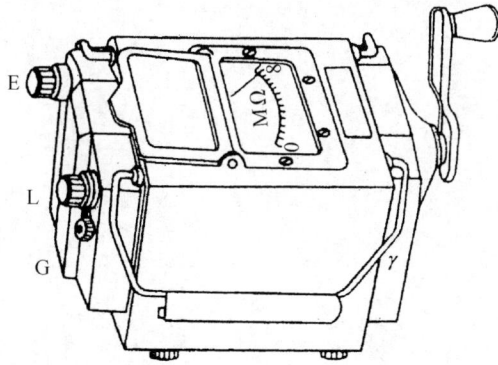

附图 3−1

兆欧表上有 3 个接线柱，分别是"线（L）"、"地（E）"和"屏蔽（G）"。测量时一般只使用兆欧表的"线（L）"和"地（E）"两个接线端接被测对象，测量线路见附图 3−2。在测量特高电阻对象。测量线路见附图 3−3。

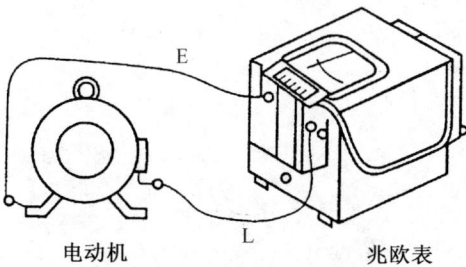

附图 3−2 兆欧表一般接法

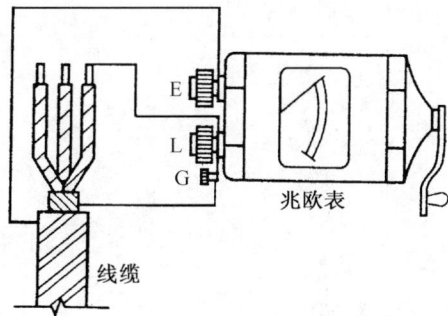

附图 3−3 需屏蔽时兆欧表接线

使用兆欧表的注意事项：

1. 测量前需使被测设备与电源完全脱离，禁止在设备带电状态下测量。

2. 测量前应先将兆欧表进行一次开路和短路试验，检查兆欧表是否正常。将兆欧表"线（L）"、"地（E）"和"屏蔽（G）"端子开路，摇动手柄应在"∞"刻度处；然后将

"线（L）"、"地（E）"短接，缓慢摇动手柄，指针应指向"0"刻度处。若不能，说明兆欧表有问题。

3．连接兆欧表与被测对象宜使用单股导线。

4．进行测量时，手柄的摇动速度尽量保持恒定，速度为大约 1208/min。待指针稳定 1min 后进行读数。

5．测量结束，应先降低手柄转速，再将被测对象对地短接放电，最后停止摇动手柄。

3　Meg-ohmmeter

Meg-ohmmeter is an instrument to measure insulated resistance. The resistance of insulated materials can be declined because of damp, heating and aging. If the resistance is out of set value, the equipment may not work normally.

A meg-ohmmeter is constructed by a high-voltage small capability generator, magnetron and measuring circuit, which also is called revolving meter. Figure A3 – 1 shows a picture of a meg-ohmmeter.

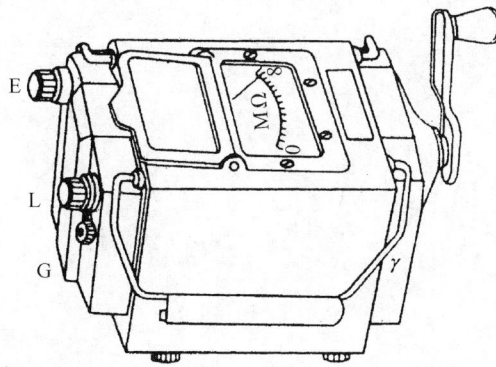

Figure A3 – 1

There are 3 terminals on a meg-ohmmeter: Line (L), Earth (E) and Shield (G). When put the meg-ohmmeter into use, there are only two terminals (L and E) being connected with object under measure. The measuring circuit is showed in Figure A3 – 2. For extreme high resistance, the measuring circuit is showed in Figure A3 – 3.

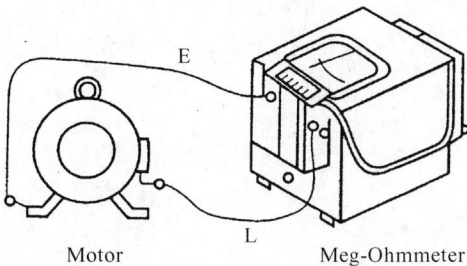

Figure A3 – 2　Connection of a Meg – Ohmmeter

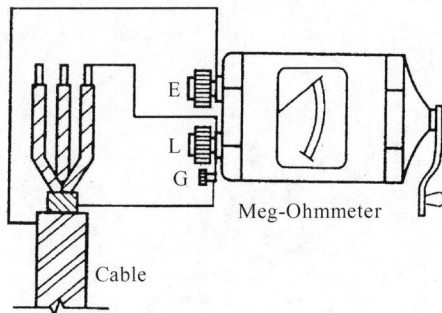

Figure A3 – 3　Connection of a Meg – Ohmmeter
with shield

Matters need attention:

1. Separate equipment with electric power before measure. It is forbidden to measure under voltage.

2. Check your meg-ohmmeter under the conditions of open-circuit and short-circuit before measure. Keep the terminals L, E and G open, while revolving the handle the pointer should stay at the scale line of " ∞ ". Then put the ends of L and E connecting together, while revolving the handle the pointer should stay at the scale line of "0".

3. The wire connecting meg-ohmmeter and equipment should be single strand.

4. While measuring the revolving speed should keep constant of about 120/min, and read value after the pointer stabilized for about 1min.

5. After reading, slow down the revolving speed, discharge the object and finally stops.

附录四　钳形电流表的使用

钳形电流表是一种可以不断开被测量电路就能直接测量交流电流的便携式仪表。正是由于钳形电流的这个特点，在电气检测等实际测量工作中使用非常灵活方便，应用相当广泛。

钳形电流表测量电流原理示意图如附图4-1所示。由图可见，被测导体中的电流为交流电流时，所产生的磁通在线圈上感应出电流，该电流通过电流表，使指针发生偏转，在钳形电流表表盘标度尺上指出被测电流值的大小。

使用钳形电流表测电流时，按动扳手，打开钳扣，将被测电流导体置于铁心内，便可以直接在电流表上读出被测线路电流的大小，使用钳形电流表应注意以下事项：

1．根据被测电流大小，合适选择量程。如果无法判定被测电流的大小，可先将量程旋转钮置于高挡，然后根据指针偏转情况，将量程旋钮调节到合适位置。

2．被测电流太小，在最低量程位置指针偏转都很小时。为了提高测量精度，可将被测载流导体在钳口的铁心上缠绕几周后再测量。此时被测电流实际值应是指示值除以缠绕圈数。

3．测量过程中，若需转换量程，只要按下扳手，张开铁心，不必切断被测电路的电源。

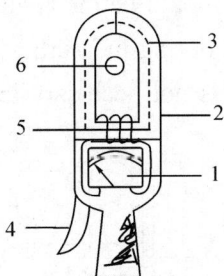

附图4-1　钳形电流表结构原理

1．电流表　2．互感器　3．铁芯　4．手柄　5．次级线圈　6．被测导线

4 Clip-on Ammeter

Clip-on ammeter is a kind of portable measuring instrument which can measure AC current directly without break circuit. Because of these features, it is widely used.

The working principle of a clip-on ammeter is shown in Figure A4 – 1. For the AC current through the wire which is being measured, it can produce magnetic flux and induce current in the circuit of the ammeter. The induced current can cause the deflection of the pointer and be measured or read.

To put the clip-on into use, press the handle to open the hook and put the wire to be measured in the center of iron core, and now the value shown on the panel can be read directly. The following matters should be pay attention:

1. Select suitable measure range. If you have no idea about the magnitude of the current under test, you can choose a sufficient high measuring range and then adjust to a suitable range.

2. If the current is so small that it cannot cause sufficient deflection of the pointer for the lowest range, you can enwind the wire on the hook for several turns to measure the current. The actual value of the current should be the result of your reading divide by the turns.

3. To switch for different range during measuring, you only need to press the handle to open the hook and turn the knob. It is not necessary to break the electric source firstly.

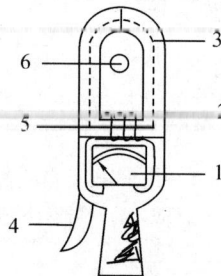

Figure A4 – 1 Clip – on Ammeter

1. Ammeter 2. Mutual Inductor 3. Iron Core 4. Handle 5. Secondary coil 6. Wire Under Test

附录五 集成运算放大器管脚分布示意图

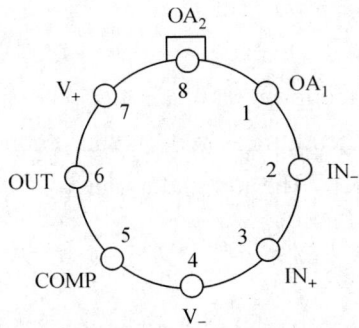

附图 5-1 F004 管脚分布示意图

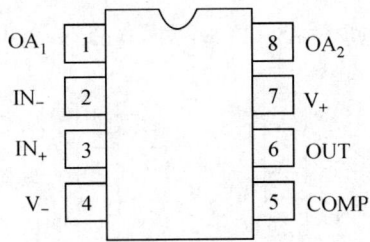

附图 5-2 OP07 管脚分布示意图

OA_1、OA_2——调零端

IN_-——反相输入端

IN_+——同相输入端

V_+——正电源

V_-——负电源

OUT——输出端

COMP——相位补偿端

5 Base Pin Array of Integrated Operational Amplifier

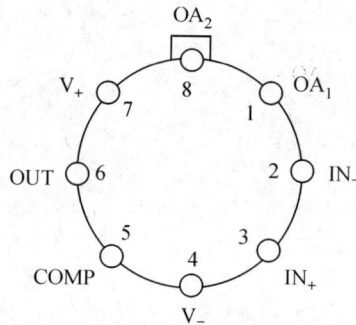

Figure A5 – 1   Base Pin Array for F004

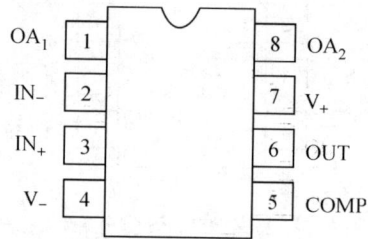

Figure A5 – 2   Base Pin Array for OP07

OA$_1$, OA$_2$——Zeroing end

IN$_-$——Inverse Phase Input End

IN$_+$——In Phase Input End

V$_+$——Positive Source End

V$_-$——Negative Source End

OUT——Output End

COMP——Phase Compensation End

附录六 集成逻辑门电路逻辑图、
逻辑表达式与外引线排列

该附录列出本讲义中用到的逻辑集成门电路和双稳态触发器的逻辑图、逻辑表达式（真值表）与外引线排列。

1. TTL 与非门

注：NC 表示空脚、GND 表示"地"

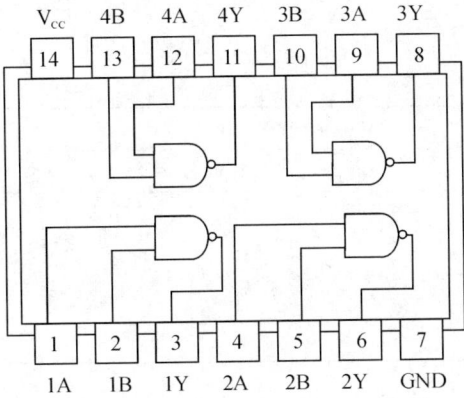

附图6-1 74LS00 四2输入与非门

$$Y = \overline{AB}$$

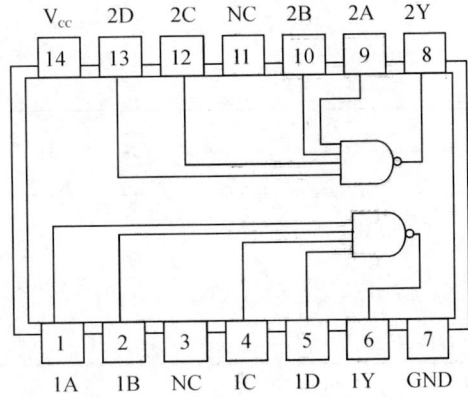

附图6-2 74LS20 双4输入与非门

$$Y = \overline{ABCD}$$

2. TTL 与或非门

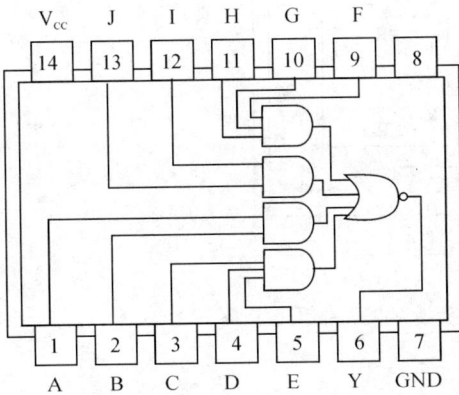

附图6-3 74LS54 2-3-3-2输入与或非门

$$Y = \overline{AB + CDE + FGH + IJ}$$

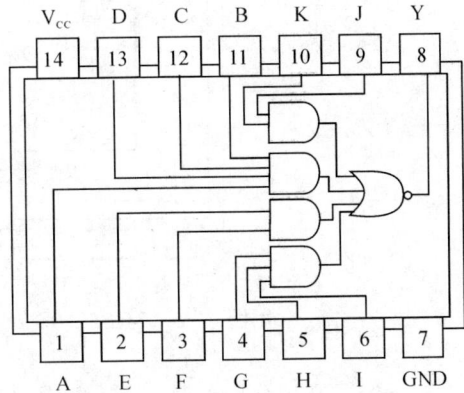

附图6-4 74LS65 4-2-3-2输入与或非门

$$Y = \overline{ABCD + EF + GHI + JK}$$

3. TTL 异或门

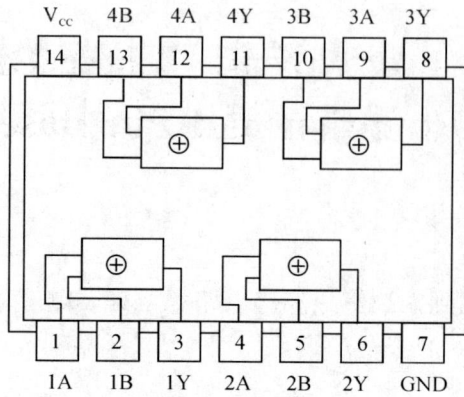

附图 6-5　74LS86 四异或门

$$Y = A \oplus B = \bar{A}B + A\bar{B}$$

输　　　　　入		输　　出
A	B	Y
L	L	L
L	H	H
H	L	H
H	H	L

L——低电平 "0"

H——高电平 "1"

4. JK 触发器

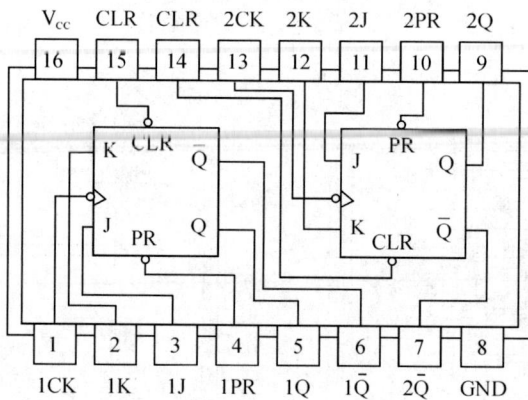

附图 6-6　74LS112 双 *JK* 触发器（负沿触发、带预置清零）

6 Logic Gate Circuit: Logic Diagram, Logic Equation and Pin Base Array

This appendix list all the logic diagrams, logic equations and their pin base arrays of the logic gate circuits which appear in this book.

1. TTL NAND gate

Note: NC denotes free end, GND denotes "Ground"

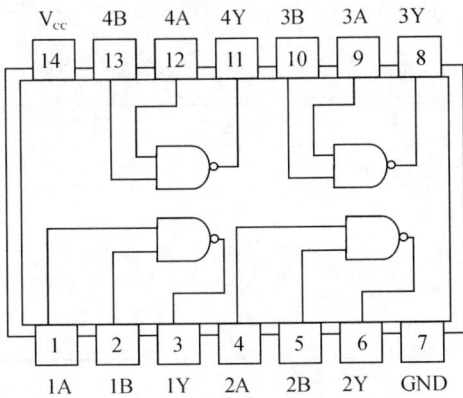

Figure A6 – 1   74LS00 4 – Double Input NAND Gate

$$Y = \overline{AB}$$

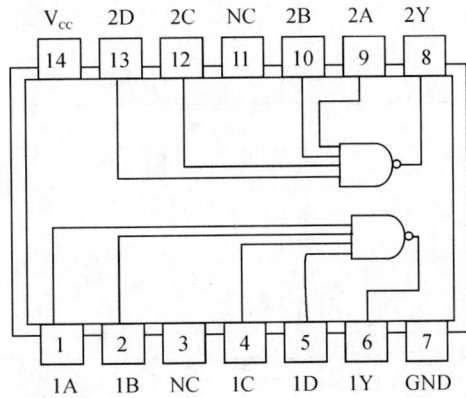

Figure A6 – 2   74LS20 2 – 4Input NAND Gate

$$Y = \overline{ABCD}$$

2. TTL AOI Gate

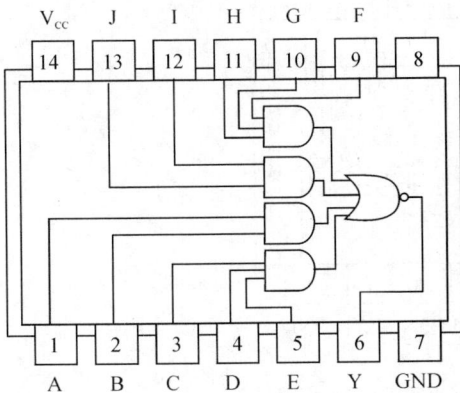

Figure A6 – 3   74LS54 2 – 3 – 3 – 2 Input AOI Gate

$$Y = \overline{AB + CDE + FGH + IJ}$$

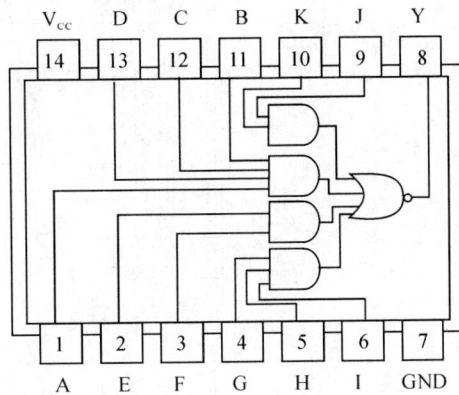

Figure A6 – 4   74LS65 4 – 2 – 3 – 2 Input AOI Gate

$$Y = \overline{ABCD + EF + GHI + JK}$$

3．TTL Exclusive-OR Gate

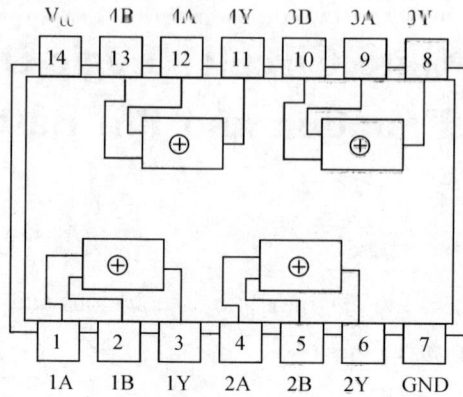

Figure A6 – 5　74LS86 4 Exclusive-OR Gate

$$Y = A \oplus B = \bar{A}B + A\bar{B}$$

Input		Output
A	B	Y
L	L	L
L	H	H
H	L	H
H	H	L

L denotes Low Voltage "0"

H denotes High Voltage "1"

4．JK Flip-flop

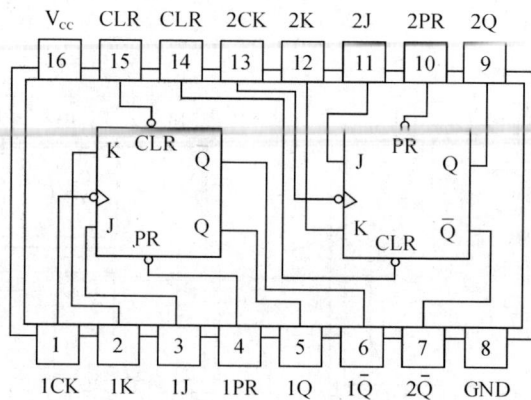

Figure A6 – 6　74LS112 Double JK Flip-flop (Fall edge
trigger, with preset zeroing button)